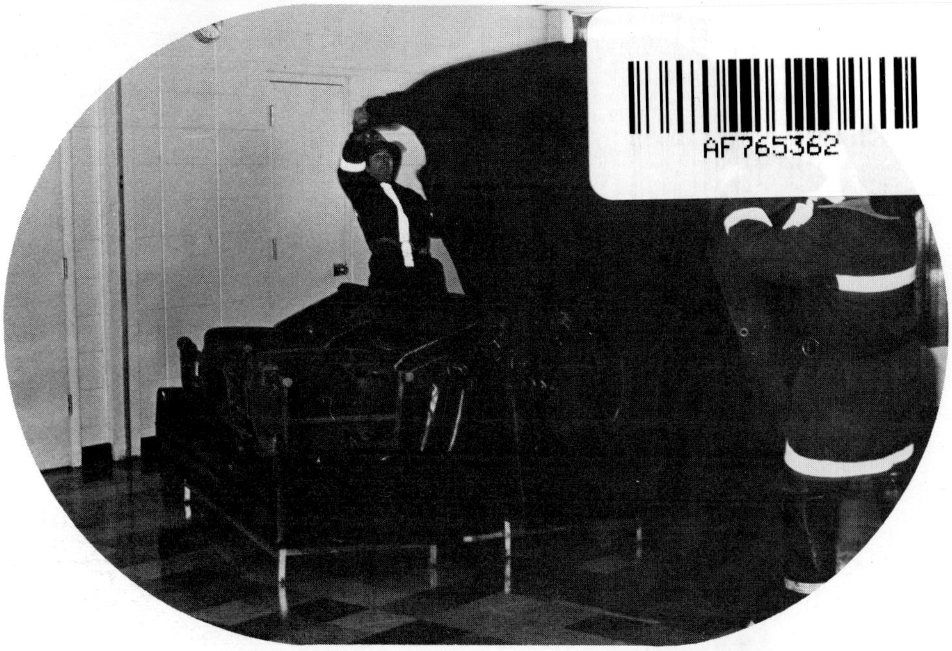

SIXTH EDITION
IFSTA 104

SALVAGE AND OVERHAUL PRACTICES

VALIDATED BY

THE INTERNATIONAL FIRE SERVICE TRAINING ASSOCIATION

PUBLISHED BY

FIRE PROTECTION PUBLICATIONS • OKLAHOMA STATE UNIVERSITY

EDITORIAL STAFF

Jerry W. Laughlin - Editor
Gene P. Carlson - Associate Editor
Connie E. Osterhout - Assistant Editor

E. T. Brown III - Photographer

PUBLISHED BY **Fire Protection Publications**
Oklahoma State University

In cooperation with
The International Association of Fire Fighters
The Insurance Services Office
Numerous State Boards for Vocational Education

Dear Firefighter:

The International Fire Service Training Association (IFSTA) is a nonprofit organization that exists for the sole purpose of serving firefighters. IFSTA is a member of the Joint Council of National Fire Service Organizations, National Fire Protection Association, and International Society of Fire Service Instructors. If you need help in locating additional information concerning the organization, training materials, or manual orders, please contact one of the people listed below.

Without you we do not exist!

Respectfully,

Jerry W. Laughlin

Write:
Customer Services
Fire Protection Publications
IFSTA Headquarters
Oklahoma State University
Stillwater, Oklahoma 74074

or call:
(405) 624-5723
Manual Orders
Training Materials

SECOND PRINTING

ISBN 0-87939-030-1
Library of Congress Number 79-84252
Sixth Edition Published 1979
Printed in the United States of America

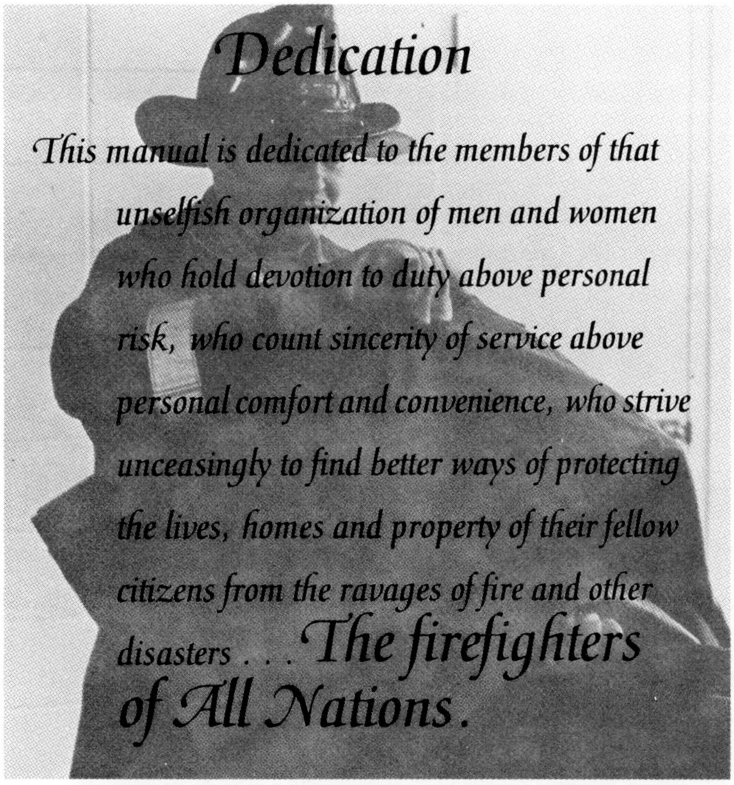

Dedication

This manual is dedicated to the members of that unselfish organization of men and women who hold devotion to duty above personal risk, who count sincerity of service above personal comfort and convenience, who strive unceasingly to find better ways of protecting the lives, homes and property of their fellow citizens from the ravages of fire and other disasters . . . The firefighters of All Nations.

Preface

As the values in property, merchandise and machinery continue to rise, efficient fireground operations to reduce losses increase in importance. An often overlooked method of saving property is the immediate initiation of appropriate salvage work. Once a fire is under control prompt overhaul following prescribed methods will also assist in loss reduction. Realizing the importance of these operations the Sixth Edition of **Salvage and Overhaul Practices** has been revised to include additional salvage operations and time saving methods.

The manual provides step-by-step techniques for performing the skills required in salvage work. It covers preparation for salvage operations, equipment used in salvage and overhaul work, methods of folding and spreading covers, improvised uses with salvage covers and handling runoff water. Overhaul procedures and operations after fire control are detailed. Information on safeguarding the property and restoring it are provided. Appendices including a suitable glossary of terms complete the manual.

The editorial staff would like to express its appreciation to the Continuing Committee, who were appointed by the IFSTA Executive Board. These men made valuable suggestions and contributions and were instrumental in the final validation of the manual.

Chairman
Ralph W. Scantlin
Arkansas Fire Academy
Little Rock, Arkansas

Secretary
Ronald Graw
Rockford Fire Department
Rockford, Illinois

Vice Chairman
Bob Baca
State Fire Marshal's Office
Santa Fe, New Mexico

A. E. Hendricks
Arkansas Fire Academy
North Little Rock, Arkansas

To the loyal and hard-working committee members who supported the Continuing Committee through numerous meetings and days of concentrated effort and who contributed to group discussion from their talent and knowledge, we also express our thanks.

Richard Calhoun
Shreveport, Louisiana

J. Humberto Guevara, S.
National Institute of Fire Prevention and Protection of Nicaragua

William Cooper
Huntington Beach, California

George Hunt
Greeneville, Tennessee

Charles Ray
Charles Ray & Associates
Albuquerque, New Mexico

Special thanks is due to the Skokie, Illinois, Fire Department under the command of Chief Jerome Burke for their assistance in obtaining many of the photographs used throughout the manual. Photos were provided by the Arkansas Fire Academy, Hale Fire Pump Company, and the Rockford, Illinois, Fire Department.

Permission was generously granted for the use of copyrighted materials by *Fire Engineering* magazine. The California State Department of Education also granted the extensive use of their materials.

Our gratitude is also extended to the following individuals whose time and talent made the final publication of this manual possible:

Carol Smith — Administrative Supervisor
Nancy Wilson — Phototypesetter Operator
Ann Moffat — Graphic Artist
Terry Bartholomew — Clerk Typist
K. Tim Perkins — Research Technician
Michael Brooks — Research Technician

NOTE: All measurements are given in English units in the text. Appendix D is a treatise on metric units and contains the text conversions referenced by chapter and page.

Contents

Introduction

Salvage work in the fire service consists of those methods and operating procedures which further reduce fire, water and smoke damage during and after fires. A portion of these damages can be attributed to the necessary operations of applying water, ventilating the building and tracing fires throughout a structure. These unavoidable procedures cannot be entirely eliminated, but improved techniques in fire extinguishment plus prompt and effective use of good salvage procedures result in a more systematic approach to minimize these losses.

Overhaul operations consist of the searching for and extinguishing hidden or remaining fires; placing the building, its contents and the fire area in a safe condition; determining the cause of the fire; and recognizing and preserving evidence of arson. Salvage and overhaul may sometimes be performed simultaneously, but overhaul generally follows salvage operations. When effective salvage procedures precede a thorough and systematic overhaul, the result will have a significant effect upon reducing the extent of the loss and facilitate prompt restoration of the property to full productive use.

PURPOSE AND SCOPE

A first introduction to salvage and overhaul work might give an impression that salvage operations are almost solely the process of spreading covers over merchandise and that overhaul is essentially a clean-up after fires. Obviously, with the preceding definitions in mind, this concept is too simplistic and erroneous. The purpose of this manual is to encourage the use of good salvage and overhaul practices in the fire service and to develop operational techniques which have been validated by the International Fire Service Training Association.

The scope of this manual covers planning operations, preparation and care of equipment, methods and techniques, salvage and overhaul practices and restoration of the premises. Firefighters must occasionally deal with situations where fires have been maliciously and intentionally started. Although firefighters may not be trained arson investigators, this manual further develops techniques of recognizing and preserving evidence of arson. To familiarize the firefighter with arson detection techniques requires: (1) his understanding that his position is one of detection rather than investigation and (2) his understanding of those things for which he is responsible.

RESPONSIBILITY OF THE FIRE DEPARTMENT

Unfortunately, some fire department officials once considered salvage and overhaul operations to be a special activity which required additional manpower. This concept can be partially understood since fire insurance companies originally financed salvage patrols in some larger cities. Experience has proven conclusively that salvage and overhaul work is closely allied with other phases of fire fighting. It is not necessary to maintain separate apparatus for this purpose; in fact, it is advisable and more practical to carry salvage and overhaul equipment on most fire apparatus in regular service.

Protect, not wreck

Present-day practice reveals that most organized fire departments have assumed the responsibility of salvage and overhaul operations. Since efficient salvage work reduces the property damage, it is the fire department's duty to perform adequate salvage. "Protect,not wreck" should be the goal. Proper and efficient salvage and overhaul work has often enabled businesses to continue after a fire with a minimum of interruption. It has also enabled families to resume living in their homes within a few hours after a fire has occurred. The search for and extinguishment of hidden fires are accepted fire department functions. With the added requirement of determining the point of origin and ignition source, especially on suspicious fires, the quality of salvage and overhaul work must be constantly improved in order to preserve fire scene evidence.

OWNER'S INTEREST AND RESPONSIBILITY

Most owners are eager to provide protection for their stock, merchandise or other valuable contents if techniques for such a program are made known to them. Good salvage procedures for owners to follow should be given by those who represent the fire service as a cooperative phase of fire department inspection surveys. These salvage procedures are allied with good maintenance and they are generally recognized as a responsibility of management. Planning and preparation should be a joint or cooperative effort between the owner and fire officials. Some of these plans and preventive measures are as follows:

- Training of employees and private fire brigades on what to do in the event of fire
- Marking and identification of:
  - Accounts receivable
  - Extinguishers
  - Floor drains
  - High value goods
  - Highly susceptible merchandise
  - Scuppers
  - Sprinklers
  - Standpipes
  - Valves
- Proper skidding of stock
- Keeping sprinkler heads free from obstructions
- Space between materials and walls
- Space between materials
- Space between materials and ceiling

This can be extended to dwelling inspections where such simple ideas as keeping all the hangers in a closet one way will gain good will and aid in salvage work.

VALUE IN PUBLIC RELATIONS

Good salvage builds good will

The damage caused by heat, smoke and water can frequently exceed the direct loss by fire, but these losses may be materially reduced through planned and well-executed salvage and overhaul practices. Fire protection people generally agree that good salvage and overhaul work is one of the most effective means of building good will. Salvage and overhaul practices must still be explained to the public even though their principles are sound. A fire department may often receive complimentary words of appreciation and praise by the news media for good salvage and overhaul operations. This praise gives the firefighters a feeling of accomplishment, particularly when the appreciation comes from people who have had their belongings saved by the firefighters. It is common for fire officials to notice better morale and efficiency among firefighters who have significantly contributed to reducing fire loss by successfully practicing salvage and overhaul. Often the little considerations are important: Protective clothing should be free of excessive grime and loose dirt (Figure 1) and firefighters should not be allowed to go charging through the premises on insignificant alarms. A good rule to build public relations is to treat the property on the fireground as if it were your own.

Figure 1 On return from a fire, protective clothing should be properly cleaned to enhance salvage on the next alarm.

DATA FOR REPORTS

Reports should tell complete story

Official reports on fires should be detailed, clear and exact. The contents of the report are confidential and should be available only to those authorized to review their contents. Official reports made concerning a fire investigate conditions that might not otherwise be brought to light. Such reports are essential on all building fires, should be made on fires that involved considerable loss and on fires that are suspicious. Some localities require reports to be made on all fires that have a loss over a certain dollar figure. Local and state laws should be checked for these requirements. An official fire department report is, in a sense, a story of conditions as they were found, what was done, one's conclusions, and departmental recommendations. Reports of this nature differ from the usual listing of fire alarms because they tell the complete story.

1 Initial Salvage Considerations

Salvage procedures, like rescue and first aid, are allied to fire fighting and require planning and preparing before they are performed. The success of any loss reduction through salvage practices depends considerably upon plans and preparation. Some of these plans have been previously mentioned with respect to responsibilities of the fire department and the owner. Additional information can be obtained from NFPA Pamphlet 604, *Recommended Practice on Salvaging Operations.* Other than preparedness through inspection surveys, the bulk of planning and preparation lies with the firefighters' training. Their success or failure can be directly attributed to the degree to which they have familiarized themselves with salvage equipment and techniques. This will also reflect on the time in which the operations are carried out. In salvage, speed counts.

COOPERATING WITH OWNER AND/OR OCCUPANT

In most instances, property owners or occupants have limited knowledge concerning salvage principles or requirements. Fire department representatives should take the initiative to inform them of recommended practices. These representatives should be prepared to train plant employees in salvage techniques in addition to giving advice on storage, plant safety and equipment. Property owners may fail to understand the limits of the firefighters' responsibility. Fire department officials can insure good relations by clarifying the limitations of their salvage responsibility with management. Best relationships will exist when fire department officials make it perfectly clear to management that salvage operations of the fire department are limited after the premises have been released to the owner. It is not the fire department's responsibility to provide manpower or equipment for post emergency operations where considerable stock must be moved. It is common for a fire department to experience difficulty in securing prompt return of salvage covers or other equipment that have been left at the property after an emergency. Management should be informed of the department's regular procedure for such usage of equipment and instructed concerning any per diem charge after a designated time.

COORDINATING PREPLANNING AND SALVAGE

The pre-fire plan should include salvage

Planning is a basic factor in all enterprises. When a particular task must be accomplished, the performance of that task requires planning. When pre-fire plans are made, all phases of fire fighting must be considered. Pre-fire planning is the process of planning fire fighting operations in advance from knowledge based on surveys of conditions and inspections of identified hazards, the relation of cause and effect, reasonable expectations, and past experiences. If a pre-fire plan is properly made, the need for salvage will be included in the plan. The need for salvage operations at fires is always present. There is usually a certain amount of water that may be sprayed upon or as runoff either comes in contact with or drips upon contents. Some factors that are pertinent to salvage operations and which must be considered in a pre-fire plan include:

- The area involved
- Nature of contents
- Merchandise requiring special handling like perishable food
- Value of contents
- Automatic sprinkler systems
- Fire protection devices
- Personnel available
- Salvage equipment available
- Ventilation devices
- Vertical openings and drains

Officers should know what to look for in an occupancy such as protecting a machine shop including covering machines to protect polished surfaces, diking and covering electric motors, covering conveyors, and removing and protecting fabric belts.

In essence salvage preplanning should consider protecting the contents against water damage by preparing to contain the water and removing it as quickly as possible to prevent further damage.

PLANNING FOR SALVAGE OPERATIONS

Operational efficiency in salvage procedures will reach its highest peak through advanced planning and training. Those responsible should review the department's salvage inventory and make certain that the kind and supply of equipment not only meets the requirements as specified in the NFPA Standard No. 1901, *Automotive Fire Apparatus,* but meets or exceeds the needs of the area protected by the department. See Appendix A for recommended salvage unit equipment. The fire department officer responsible for training should see that all firefighters are adequately trained in salvage operations and in the use of salvage equipment. Fire officers should be instructed to consider salvage potentials and not hesitate to immediately call additional help for salvage operations.

Planning includes equipment inventory and training

SIZE-UP

Efficient fire department operations are based upon certain definite principles, and one of the principles of fire service work is that a size-up must be made to gather data for decisions on what to do and how to do it. Officers responsible for salvage operations should develop a system for such operations. Salvage size-up should begin immediately on receipt of the alarm considering the type of building, the occupancy and contents. One effective system of size-up includes the following steps:

- Facts
- Probabilities
- Resources available
- Decision
- Plan of operation

While this system is designed for overall fireground operations, it can readily be adapted to salvage.

Facts to be determined include the nature of the emergency, the size of the fire and its location within the structure, the type of construction in the involved building, the type of contents exposed, the fire protection features of the building, the weather, and anything else regarding the building and its contents that could aid or hinder salvage operations. When floors are checked above and below the fire, the salvage needs should be determined. Operations should begin on the floor below the fire. The majority of these facts must be gathered through pre-fire planning surveys.

Consider salvage in size-up

Probabilities must include an estimate of where the fire may travel, preventable damage, possible hazards to personnel engaged in salvage, effects of possible weather changes upon the operations under consideration and any other probability relating to operations.

Resources available to deal with the emergency are such items as apparatus, equipment and personnel. Officers assigned to salvage must be adaptable. They must have the ability to adapt their apparatus and equipment to various situations, and they must be able to utilize equipment and facilities that may be available on the fire premises or available from other sources. Lack of personnel is almost always a problem. A good salvage officer will develop an organization that will be effective at the outset with a minimum number of personnel. His nucleus will be readily expandable in order to use additional personnel that may become available (1) as the fire is brought under control; (2) through greater alarms or special calls; or (3) through the use of the civilians employed at the occupancy where a fire occurs.

The **decision** step essentially involves what to do and how to do it based upon an evaluation of the facts, probabilities and resources available. Fireground activities are not static; fireground situations can change very rapidly and decisions may have to be re-evaluated. Therefore, a great number of decisions are required, and the officer in charge of salvage must be ready to make those decisions. He must be available to receive information and to communicate his decisions, which means that his subordinates must be able to locate him. A modified command-post operation may be required for large salvage operations.

What to do and how to do it

A **plan of operation** which results from size-up is a mental diagram of the processes necessary to achieve the objectives. The plan should be as simple as possible so as to be readily communicated and easily understood. After the plan is activated, constant supervision is required so that the plan will be consistently directed toward the objectives.

OTHER CONSIDERATIONS

All fire service personnel should be proficient in forcible entry so that they know where and how to gain access to a structure with a minimum of damage. The ideal of simply turning a knob is not present in all situations; frequently the entry is locked or otherwise obstructed, and the firefighter must use special tools. Salvage begins with care in forcing entry and the clearing of windows so that frames are not unnecessarily damaged. Watch for heavy flame conditions in ceiling areas that indicate the possible dropping of the ceiling and the need to remove the contents of the room.

Salvage operations begin with forcing entry

As a salvage operation, ventilation is essential if damage from fire, smoke and heat is to be prevented. Rapid removal of smoke, heat and gases during the early stages of fire pinpoints the area involved, allows firefighters to get closer to the source of the fire and provides a cleaner atmosphere in which to work. Ventilation also minimizes the use of indiscriminate streams of water in areas where contents can be damaged or broken.

The most significant of all salvage principles may be the accurate application of water. Applying water at the wrong time, from the wrong place, with the wrong nozzle pattern and in unnecessary amounts causes a great deal of preventable damage. Water should not be applied in smoke-filled areas until ventilation has been accomplished and the seat of the fire can be found. The wrong place may be either inside or outside of structures. Although high-velocity, variable-pattern fog nozzles provide a means of using water with the greatest degree of efficiency, nozzles can contribute to unnecessary losses when used incorrectly.

Water is delivered not only at a high velocity, but in fine droplets that readily convert to steam. This can result in great volumes of flame, heat and smoke being forced away from the point of application. Therefore, nozzlemen must be trained in the correct use of fire streams. Applying water from the outside of a structure should be avoided in every possible instance. This may also force flame, heat and smoke into uninvolved sections of the structure, sections that could contain articles highly susceptible to heat and smoke damage.

A similar situation may occur where water is applied incorrectly within a structure. Ideally, water should be applied from a position that will force fire gases and heat to the exterior through windows, doors or other openings. Application of water from the correct location will not only reduce unnecessary damage, but will also control a fire more quickly and serve to ventilate the fire area. A major factor in reducing water damage is the immediate closing of a nozzle once the fire has been knocked down. If additional water is applied, it will not be converted to steam and will do additional damage as it runs off.

Correct water application is essential to salvage

Indiscriminate application of large quantities of water on stocks of paper goods, baled paper or fabric, grain and other bulk or porous materials will cause excessive water damage and could cause overloading of floors with subsequent collapse. This, in turn, could result in injury or death to personnel or unnecessary damage to the building and contents.

ARRANGING MERCHANDISE TO BE COVERED

The actual arranging of merchandise to be covered can be limited when large stocks and display features are involved. Display shelves are frequently built to the ceiling and directly against the wall. This construction feature makes it difficult to cover shelving. When water flows down a wall it will naturally come in contact with each shelf and wet the merchandise. A better construction feature would be to allow adequate clearance between the shelves and the wall. This construction feature should be discussed with the owner/occupant during a preplan inspection.

One common obstacle to efficient salvage work is the lack of skids under all stock which is susceptible to water damage. Some examples of merchandise which have perishable characteristics are: flour, material in cardboard boxes, feed, paper and other dry goods. Stock in basements should be placed on skids at least five inches above the floor, but stock on upper floors is reasonably safe with skids of lesser heights. When salvage covers are limited, it is advisable to cover the more valuable merchandise and sacrifice the less valuable. It is also good practice, when there is a shortage of covers, to use available covers for water chutes and catchalls even though the water must be routed to the floor and cleaned up afterward.

Merchandise stacked too close to the ceiling will also present salvage problems. There should be enough space between the stock and ceiling to allow men to easily apply covers. Material lying around on floors makes salvage work much more difficult and the employees of an operation should be cautioned to arrange all material in neat and orderly stacks. Firefighters should be aware of the more costly and essential contents of a building and make every effort to protect these items first.

Expensive machinery may require special salvage attention, but the high values of intricate equipment would make an extra effort worthwhile. Pre-fire inspections allow for planning to establish which machines receive top priorities for covering. If this is not done before a fire, it may be possible to get some of the information from evacuated employees. Where valuable equipment may be easily protected from water damage but still subject to smoke or heat damage, ventilating procedures must be attended as soon as possible.

Protect important records and documents

Where fire involves a business or a manufacturing concern, a sincere effort must be made to protect the records of the occupant. The important point is to save insurance policies, stocks and bonds, orders, inventories, accounts receivable, patents, designs and other records of business transactions. American Insurance Association reports show that only a small percent of businesses are able to reopen after their records have been destroyed by fire; therefore, the importance of this phase of salvage work must be stressed repeatedly in training sessions.

ARRANGING HOUSEHOLD FURNISHINGS TO BE COVERED

Household furnishings present a different type situation than does merchandise. If a reasonable degree of care is taken, one average size cover will usually protect the contents of one room. When circumstances limit the covering of all items, those of greater value should be covered first. One good method of arranging furnishings in the average bedroom is illustrated in Figure 1-1.

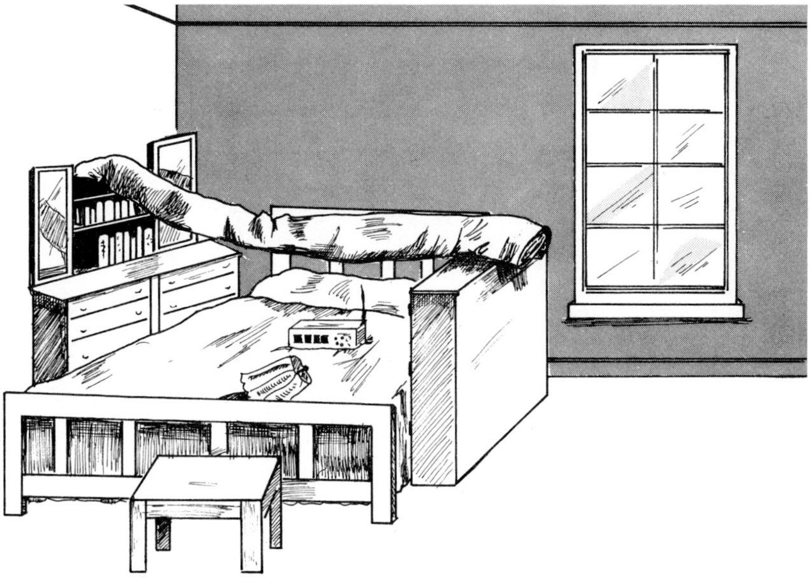

Figure 1-1 When arranging furniture, place those pieces with drawers facing toward the center of the arrangement. This will reduce the possibility of water leakage into the contents of the drawers.

A suggested procedure for arranging furniture in a room is as follows: group the furniture in the center of the room, but not under a light fixture which might leak. If the floor covering is a removable rug, slip the rug from under the furniture as each piece is moved and roll it for convenience. A dresser, chest or high object may be placed at the end of the bed. If there is a roll of rug, place it on top to serve as a ridge pole. Other furniture can be grouped close by and pictures, curtains, lamps and clothing can be placed upon the bed.

Valuables should be placed in drawers with the drawers placed inward or removed. Highly valuable items should be given to an officer for proper disposition. No matter the intent, firefighters should never place valuables in their pockets.

It may sometimes be necessary to place the cover into position before some articles are placed on the bed. In this event, bed and furniture can be protected while other items are placed under the cover. Due care should be taken not to scratch or mar the finish or furniture since this could be worse than water damage. Figure 1-2 illustrates a method of arranging living room furniture.

Figure 1-2 The sofa is an excellent location to place pictures and large knickknacks. Try to make the arrangement slope outward with the taller objects in the center.

Contents of residences can be protected by one or two men well trained in salvage work. This operation begins in the room directly under or adjacent to the fire area and involves grouping the contents so that a minimum number of covers will be required. The salvage personnel should observe the following rules:

- Do not overlook the contents of closets.
- Check the tops of bureaus, dressers and tables, and place small valuables found on these pieces in drawers for safekeeping.
- Use the largest and heaviest pieces of furniture as the base for the grouping. Before other articles are piled on them, protect fabric coverings and wood finishes with rugs, blankets, cushions or pillows.
- Do not build the pile too close to a wall where spreading covers will be difficult.
- Roll up any rugs, if possible, and place them on top of the grouping. If a rug cannot be removed, try to roll it from both ends toward the pile.
- Do not overlook paintings or wall tapestries, which may be placed on the pile after the larger and heavier pieces of furniture have been grouped.
- After salvage covers have been placed, tuck the edges around the base of the grouping and the rolled-up rug.

Where the floor or floor covering should be protected in addition to the grouped furnishings, covers can be placed in the order shown in Figure 1-3.

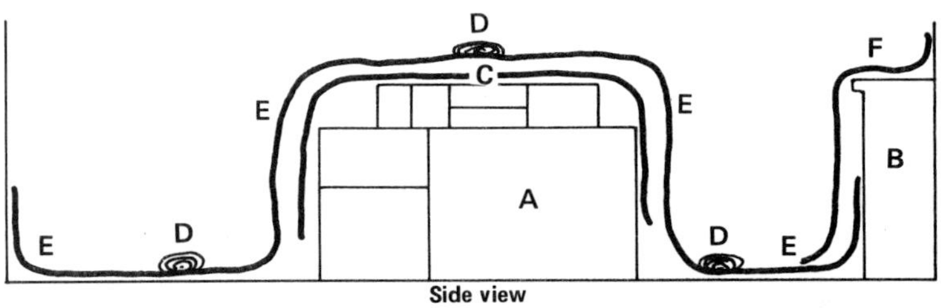

Figure 1-3 In occupancies with wall-to-wall carpeting it will be necessary to place the covers as illustrated to provide complete floor coverage and minimum water damage.

OUTSIDE STORAGE

At times it may be necessary to protect materials stored outside from excessive water damage. This may require covering with tarps or building dikes to channel runoff. The opportunity to perform salvage and save additional property from damage should be included in the size-up of every fire.

A great deal of salvage can be accomplished by a good officer and his crew if they will think salvage and utilize their training coupled with applying a lot of common sense.

2

Salvage Equipment: Care and Preparation

Fire department officials may select various kinds of salvage equipment to be used in their salvage activities, but local conditions will ordinarily determine the equipment that is needed. Although there are many salvage tools and accessories which make salvage operations more successful, salvage covers are considered to be the basic salvage equipment. For this reason, salvage covers will be studied first and a study of equipment other than covers will be discussed by referring to suggested kits. Considering equipment in this manner will associate the equipment with the job to be done.

SALVAGE COVERS

The extent to which a fire department can perform salvage cover work depends largely upon the number and size of covers available. Conventional salvage covers are made from closely woven canvas materials such as cotton duck which have been treated for waterproofing and fire resistance. Salvage covers of rubber-coated cotton duck have been used and plastic covers are also becoming popular. Disposable plastic salvage covers are also in limited use. See Appendix B for additional information on disposable covers. Salvage covers have reinforced corners and edge hems into which grommets are placed for hanging or draping the covers.

Salvage covers can be obtained in various sizes, but the sizes ranging from 9 by 12 feet to 14 by 18 feet are most common. The smaller sizes lend themselves to easy handling, but the larger sizes will cover more merchandise or furnishings.

A relatively new trend, the use of plastic salvage covers, shows some promising advantages. These covers are made from 100 percent polyethylene. This type of cover is extremely lightweight and is easily handled. A cover made from No. 3 heavyweight polyethylene and comparable in size to canvas covers weighs about one-tenth, so the weight of the plastic covers is definitely advantageous. These covers are quite economical and are practical for indoor and outdoor use. They are chemically inert and are not generally affected nor penetrated by alkalies, oils, acids, caustics or solvents. They are relatively unaffected by normal temperatures, will remain flexible below zero degrees, and will not mold, mildew or absorb moisture. For better visibility, these covers can be obtained in many colors. Plastic covers do, however, have a tendency to slip from highly piled merchandise and are not well adapted to bagging by rolling the edges. Plastic covers should not be utilized in areas where sparks or hot embers may be present because the cover will be easily penetrated.

Plastic covers resist chemicals and temperature changes, but can be damaged by embers

TESTING SALVAGE COVERS

When a new shipment of salvage covers is received, one cover should be selected at random for a test known as the "lake test." Salvage covers which have been in service and about which there is doubt concerning their protective qualities may also be tested by the same method. The lake test procedure is as follows:

- Spread the cover out flat on a smooth surface (floor or driveway).
- Either roll the edges of the cover as described on page 69 under the Catchall or build a dike with 2- by 6-inch lumber around all edges. (The lumber dike provides a better test.)
- Fill the basin with water, preferably to a depth of approximately five inches.
- Permit this arrangement to stand for at least 12 hours with the cover resting on the floor. After the water is removed there should be no signs of excess leakage on the underside. Leakage should not be confused with normal bleeding or sweating.

MAINTAINING SALVAGE COVERS

A good supply of salvage covers represents a considerable investment, and their care and preparation are very important. With proper cleaning, drying, repairing and general care their span of service can be considerably extended. Preparing salvage covers for use involves all the above-mentioned functions before they are placed into service.

Cleaning

Ordinarily, the only cleaning that is required for salvage covers is flushing with a hose stream and scrubbing with a broom as shown in Figure 2-1. This cleaning process should be done on a clean floor or concrete, and the cover should be thoroughly rinsed after it is hung to dry. Covers that are extremely dirty and stained may be scrubbed with a mild detergent solution and then thoroughly rinsed. Permitting salvage covers to dry while in a dirty condition is not a good practice because after carbon and ash stains have dried they are difficult to remove even with a detergent. Wood ash and other products of burned materials may also have a chemical effect upon the fabric of the cover. If it is necessary to postpone the cleaning of dirty covers because of other important activities, it is advisable to wet down the dirty covers so they will remain damp until they can be cleaned.

Figure 2-1 Salvage covers should never be allowed to dry before being cleaned. Wood ash and other products of burned materials may have a chemical effect on the cover.

Figure 2-2 Salvage cover drying racks should allow for adequate drainage and air circulation.

Drying

Salvage covers should be thoroughly dry before they are folded and placed in service. This practice is essential in order to prevent mildew and rot. A well-ventilated place should be provided for drying and the drying area should be somewhat isolated (Figure 2-2). Salvage cover drying racks are sometimes provided at the rear of an apparatus floor. The objection to outdoor drying of salvage covers is that the wind tends to blow and whip the covers, producing excessive stress upon the covers. Outdoor drying racks that will withstand this whipping action caused by the wind are difficult to construct.

Examining

After salvage covers are dried, they should be examined for damage which they may have received (Figure 2-3). One of the more efficient, quick methods of examining covers is described. The cover is arranged in a line on the floor and three or four firefighters take their positions alongside. The cover is raised up and over the heads of the men and each man watches for holes and torn places as the cover is moved slowly overhead. When the cover is well overhead, the watchers are in semidarkness and any hole in the cover will show up clearly by the light filtering through from the opposite side. When a hole is found, a watcher holds the palm of his hand against the cover and another man on the outside marks the hole with chalk.

Figure 2-3 Examine the cover for holes and tears. If any are found, they should be marked with chalk and repaired immediately. A good method to search for holes, as shown here, is to use a light on one side to illuminate holes on the opposite side.

Repairing

The developments made in filament or plastic adhesive tapes have changed the entire practice of mending covers. It is no longer necessary to hand sew small holes and tears or patch them with various types of cement or glue.

Any small hole or tear can now be easily mended with a tape which adheres permanently to most materials. There are also iron-on patches available. Some departments still prefer to machine sew large tears and patches and then cover the stitches with a mending adhesive tape. The special adhesive tape should be applied on a clean, dry surface for best results. If the mending tape is applied to both sides of the surface, it will increase the bond for a stronger and more permanent seal. This tape is waterproof and also stands up under normal heat or cold. Polyethylene covers can also be mended with a plastic film tape without an adverse effect upon the waterproof qualities (Figure 2-4).

Figure 2-4 Small holes and tears are repaired with adhesive tape. Place the tape on clean, dry surfaces on both sides of the cover for a strong permanent seal.

Large tears or holes may also be effectively repaired by using pieces of canvas and contact cement. Some firefighters, who have had experience both ways, feel that this method is more durable and permanent than the adhesive tape method. The following steps are a good recommended procedure to follow:

- Use canvas patches and contact cement.
- Extend the patch at least two inches beyond the hole or rip to insure a strong waterproof repair as illustrated in Figure 2-5.
- Patch only one side of the simple puncture or a small rip.
- Patch both sides of the cover for larger holes up to three inches in diameter and longer rips. Holes more than three inches in diameter and very long rips are difficult to repair; they usually require the service of professional repairshops such as a tent and awning company or a shoe repair service. The best way to repair covers is with a close "herringbone" stitch. The thread and sewn area should then be waterproofed. When patches are required, they should be sewn with double seams and waterproofed.
- Place tape over any stitching to insure a waterproof seal before returning the cover to service.

Figure 2-5 Large holes and tears can be repaired by using a canvas patch and contact cement or if preferred the patch may be sewn.

Placing In Service

Salvage covers that have been properly cleaned, dried, examined and repaired are now ready to be folded and placed in service or in stock. Most departments carry all their covers on the apparatus, but some departments have a supply in stock to replace soiled or damaged covers during the processing time (Figure 2-6). The type of salvage cover fold may be selected from the methods shown in this manual. The covers should be located on the apparatus so that they will be readily accessible to firefighters and protected against injury from other fire department equipment (Figure 2-7).

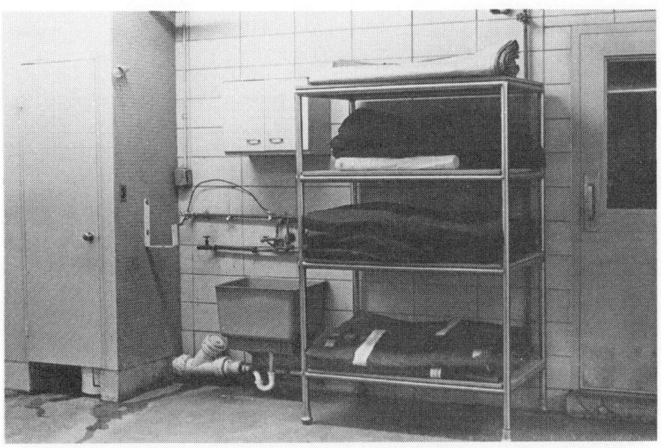

Figure 2-6 Extra salvage covers should be stored on shelves in a clean, dry, well-ventilated area. Roof covers can be marked for easy identification such as those on the lower shelf. Note also the large laundry sink mounted low on the wall for washing boots before removing them.

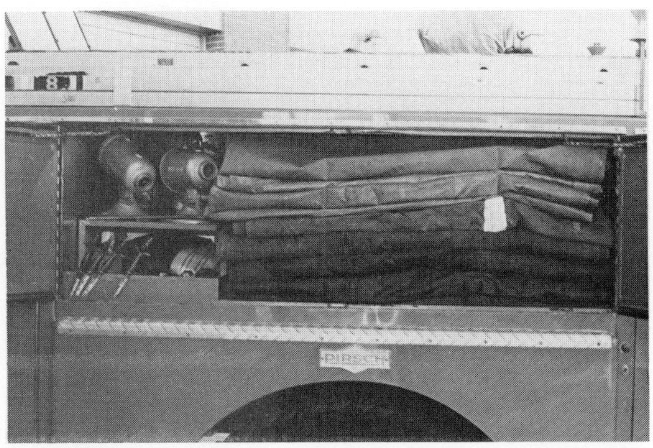

Figure 2-7 Covers stored on apparatus out of the weather and ready for use.

Practically all salvage equipment needs care and upkeep, especially salvage covers which are more susceptible to damage. Every fire department should have a set of rules and regulations governing the care of covers. A set of rules is submitted here which can be revised or added to as a department may desire.

Establish departmental policy for salvage covers

- For records and identification all covers should be adequately marked. Inventory, service and maintenance records should be kept.
- Throwing covers from roofs, windows or other openings above the ground is strictly prohibited.
- Unnecessary walking on covers should be prohibited, and those covering floors and stairs should be picked up promptly to avoid damage. Frozen covers should not be walked on to force folding; rather they should be thawed, cleaned and dried before folding.
- When men are moving covers over sharp projections or from stock or machinery, care must be used to prevent tearing.
- Covers must be washed with soap and water when necessary, and hung on the drying racks with air circulation on both sides as soon as possible after returning from fires to prevent mildew. Covers must be thoroughly dry before folding for storage.
- After covers are dry, they must be examined for holes and mended before they are refolded and placed on the apparatus.
- Covers should be taken from the apparatus and refolded according to department policies.
- A receipt must be obtained for covers left at an incident. Any policies and charges should be outlined when the covers are left.

A general rule for using covers is if there is any doubt as to their need, cover the object. Any materials that are already wet are usually passed over so that available covers can be utilized to protect dry items.

SALVAGE KITS

For convenience during salvage work at fires, it is suggested that salvage equipment be arranged in kits. Each kit should be supplied with the selected tools and accessories that are necessary to effectively perform a specific job. With this arrangement, it is not necessary to hunt for equipment on the apparatus when a specific job needs to be done. Some desirable salvage kits are discussed and described in the following paragraphs.

Metalworking Tool Kit

A metalworking tool kit, containing the following suggested tools, permits a salvage crew to perform many special jobs that could not otherwise be accomplished.

Electricians' pliers
Chisel
Crescent wrench
Hammers
Hacksaw
Pipe wrenches
Applicable power tools
Hydraulic jack
Side cutter
Screwdriver and pliers

Automatic Sprinkler Kit

The tools and supplies for an automatic sprinkler kit are used primarily to control a flow of water from an open sprinkler head. An open sprinkler head can do considerable damage to merchandise after a fire has been controlled. It is not usually advisable to close any water supply control valves to the sprinkler system because this action may eliminate or nullify sprinkler protection in other areas where it might be needed should a fire occur. Suggested tools to be included in the sprinkler kit are listed and some of these are discussed following the list: (NFPA Standard 1901 recommendations.)

Sprinkler head wrenches
Assortment of caps and plugs
6 sprinkler stoppers, tongs or wedges
Special bent wrenches to reach obscure valves
Pipe wrenches
12 standard sprinkler heads (assorted temperatures and types)

Figure 2-8 Shown here are several examples of sprinkler head tongs and wedges. These tools are used to help reduce water damage from an open head, rather than shutting the entire system down. The wedges are constructed of soft wood and sometimes covered with rubber to help seal the orifice. As the name implies, the wedge is forced into the space between the orifice and the frame. The tongs are a mechanical device that uses leverage to seal the orifice and stop the flow of water.

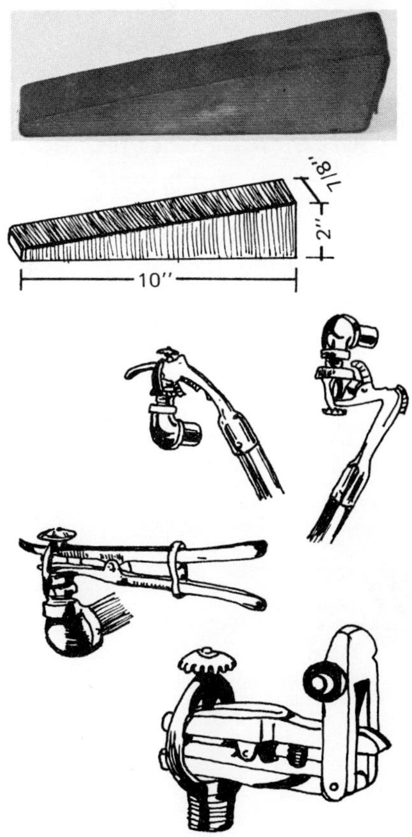

Various sizes of pipe caps and plugs may be carried in a sprinkler kit. A sprinkler pipe may freeze and burst or break from other causes. A fire department is permitted to make necessary temporary repairs. For this operation, the system must be shut off and then reopened after repairs. Since it is not advisable to close the main sprinkler valve unless all traces of a fire are absolutely extinguished, sprinkler tongs and wedges should be used to shut off the flow of water at the operating heads. Several types of sprinkler head tongs and wedges, as shown in Figure 2-8, should be carried in the sprinkler kit. The wedge is made of soft wood which can be forced against some sprinkler openings to shut off the water. Sprinkler tongs work on the clamp or leverage principles as illustrated. Sprinkler heads in a deluge system are not capped with a fusible link, so the supply valve which controls the deluge system must be closed to shut off the water. This supply valve is not the main control valve.

Mop-Up Kit

This kit is to be used for cleaning floor areas after heavy debris has been removed. The equipment suggested here represents a minimum. An additional supply may be added.

Two mop handles and detachable mop heads
Two rubber squeegees with rubber in good condition
Water scoops
Two corn brooms
Mop wringer with bucket
Water vacuum
Submersible pump
Auger with reverse taper or
Electric drill with cutters up to six inches in diameter

Clean-Up Kit

Clean-up kits differ from mop-up kits in that they are to be used for definite articles such as furnishings, walls and clothing. Some of the suggested equipment to be carried in this kit is listed, and represents a minimum.

Kits include selected tools for specific jobs

Four sponges
Two chamois
Bags of sawdust
Two scrub brushes
One whisk broom
Assortment of rags
Two buckets
Deodorizer
Scoop shovel

Woodworking Tool Kit

In order to accomplish complete salvage work at fires, it is sometimes necessary to rearrange or construct certain facilities for water chutes, bracing or temporary repairs to the building. If a woodworking tool kit is provided on the apparatus, such construction work can be done without delay. The tools suggested for such a kit are as follows:

One crosscut handsaw
One claw hammer
An assortment of nails
An assortment of screws
Tarpaper or plastic sheeting
Wood lath
Roof covering
One wrecking bar
One padlock and hasp
Applicable power tools
Screwdrivers
Heavy-duty stapler

ACCESSIBLE SPRINKLER EQUIPMENT

It is not practical to carry a large assortment of sprinkler heads in a fire department sprinkler kit because a sprinkler head cabinet, similar to the one shown in Figure 2-9, is usually located near the sprinkler valve installation. It is the fire department's responsibility to control the water flow and to see that the system is restored by those who are responsible for the building. Special wrenches are occasionally needed when working on sprinkler valves (Figure 2-10).

Although firefighters do not restore sprinkler systems they may be required to remove sprinkler heads. The sprinkler cabinet should contain a sprinkler head wrench, but it is a good policy to carry one or two sprinkler wrenches in the sprinkler kit. Ordinary wrenches are not well adapted to fit the sprinkler heads since the jaws of ordinary wrenches are not deep enough to provide a good grip. How this type wrench fits a sprinkler head is illustrated in Figure 2-11.

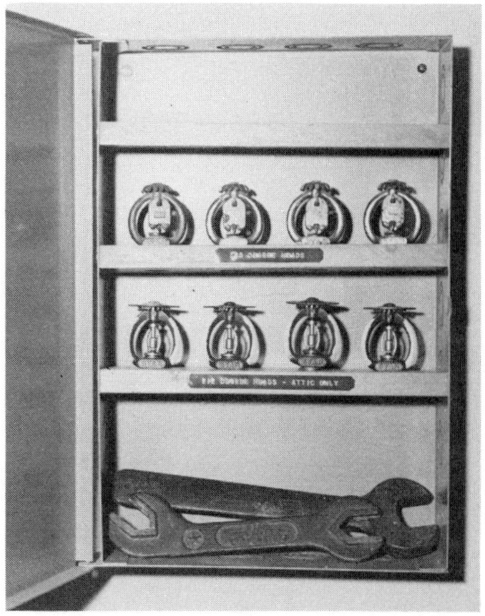

Figure 2-9 Spare sprinkler heads and wrenches should be available on the premises.

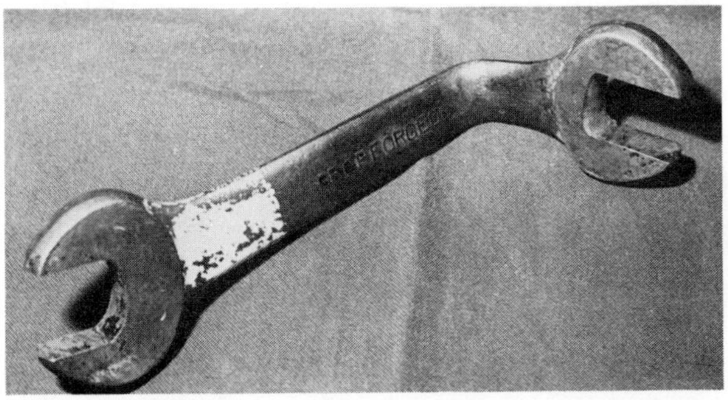

Figure 2-10 Special wrenches may be necessary to work on sprinkler valves placed in tight locations.

Figure 2-11 Sprinkler head wrenches which have deeper jaws to fit sprinkler heads should be carried by the fire department.

OTHER SALVAGE TOOLS AND EQUIPMENT

The tools and accessories which are recommended for salvage kits are associated with particular jobs. Other tools which may be required include mauls, star drills or concrete cutting saws for draining masonry floors. Forcible entry tools will be used in gaining access to locked areas or floors to perform salvage. Reference IFSTA 101, **Forcible Entry, Rope and Portable Extinguisher Practices.** Other salvage equipment consists of devices or machines that perform specific functions other than those which are performed by salvage covers. These other salvage devices supplement the use of covers, aid in good salvage practices, and further develop good public relations. Portable generators to supply emergency lighting enhance salvage work at night or in windowless buildings.

Water Chutes

Water chutes are devices to route water short distances through either doors, windows or other openings. Usually, these chutes are either made from an old or discarded salvage cover or they are constructed by improvised methods at fires. Commercial water chutes are not common, but are available. Some methods for making water chutes are discussed under Constructing Water Chutes on page 74.

Floor Runners

Figure 2-12 Floor runners will prevent additional damage to floor surfaces or coverings from debris, water or materials tracked in.

Floor runners are usually made from old salvage covers; however, canvas can be bought for this purpose. Runners should be about three feet wide and 20 to 30 feet long. They should be rolled and tied with a quick-release knot and carried in a convenient location on the apparatus. A recommended location is with the officer in the cab so he can extend it upon entering the premises for size-up. One of the best ways to develop public good will and also prevent needless damage to floors and/or floor coverings is to use a prepared floor runner. This runner can be unrolled from an entrance to most any part of an ordinary home area. Costly floor coverings are sometimes damaged by mud and grime which may be tracked in by firefighters when a floor runner is not used. A floor runner that has been applied is shown in Figure 2-12. Floor runners should never be applied over a floor furnace unless it is certain that the burner and pilot have been turned off. Vinyl-laminated nylon floor runners are commercially prepared and are lightweight, flexible, tough, heat and water resistant, and easy to maintain. When floor runners are removed after operations are completed, care must be taken that debris is not dropped. The runner should be folded lengthwise about 8 or 10 inches in from the sides and then loose folded or rolled from each end toward the middle.

Bags of Sawdust

Sawdust may be used in salvage practices to take up moisture from floors and to build dikes to route water out of buildings. This practice will help to keep floors from warping if the wet sawdust is not left too long. The kind of sawdust best suited for this purpose may be obtained from any cabinet shop where white pine is used. White pine sawdust is preferred because it will hold together when wet and because it will not stain a finished floor like many of the hardwood sawdusts. For the best results, the white pine sawdust should be screened before use in salvage activities.

Portable Pumps

Portable water pumps are used by many fire departments after fires to remove water from basements, elevator shafts and sumps. Fire department pumpers should never be used for this purpose because fire pumps are intricate and expensive machines and are not intended to pump the dirty, gritty water found in such places. Diaphragm type portables and submersible pumps are also useful in dewatering.

Various other kinds of devices known as eductors, ejectors, jets or siphons may be used for the removal of excess water from basements, cellars, vaults, manholes and other confined areas. These devices can be moved to any point where a line of hose can be placed and an outlet for water can be provided. Equipment of this nature will be discussed and illustrated on page 94.

"S" Hooks

Salvage covers may be secured with "S" hooks or other materials. Salvage cover "S" hooks resemble meat hooks. In fact, they may be purchased from meat-handling equipment suppliers as well as made in fire department shops. "S" hooks should be correctly shaped and sturdy enough to be driven into any type of wall including masonry. Fire department made "S" hooks are usually constructed from 1/4-inch or 5/16-inch cold-rolled steel. The larger size is recommended because it is more durable. Stock is cut into 8-inch lengths, shaped on an anvil and sharpened on both ends (See Figure 3-94, page 62).

Ike-O-Hook

The Ike-O-Hook is a versatile handcrafted device used in salvage and overhaul operations. It is constructed of materials as used for the "S" hook. The eyelet as shown in Figure 2-13 can be placed over the point of an upright pike pole allowing a salvage cover to hang from the hook. This procedure can be utilized in forming smoke curtains or water chutes. The eyelet of the Ike-O-Hook can also be secured in place by hose straps. By placing the hook through the grommet and pulling it tight, a bucket of water, or any convenient weight that could be hung from the hook can be used to weight down a salvage cover. This operation is shown in Figure 2-14 in conjunction with a water chute. The Ike-O-Hook

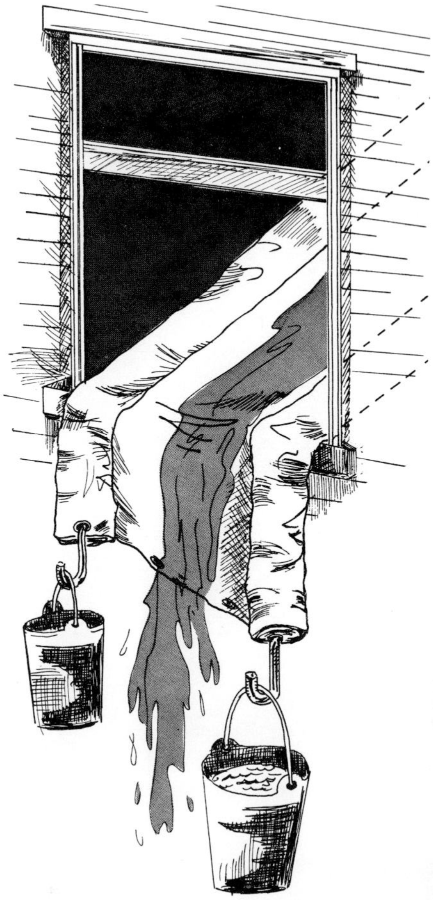

Figure 2-14 A combination of Ike-O-Hooks and weights may be used with a water chute to hold it in place and keep it taut for proper drainage.

Figure 2-13 Ike-O-Hooks for use in salvage operations.

can also be used to elevate an extension cord off the floor by securing the eyelet and placing the extension cord over the hook (Figure 2-15). The use of the Ike-O-Hook is only limited by the imagination and creativity of the user. This device performs the same as the U-shaped pike pole hook that was formerly carried on many ladder and salvage companies for lifting covers over cabinets, high shelves or equipment (Figure 2-16).

Figure 2-15 The Ike-O-Hooks can be used with pike poles to keep extension cords out of pooling water.

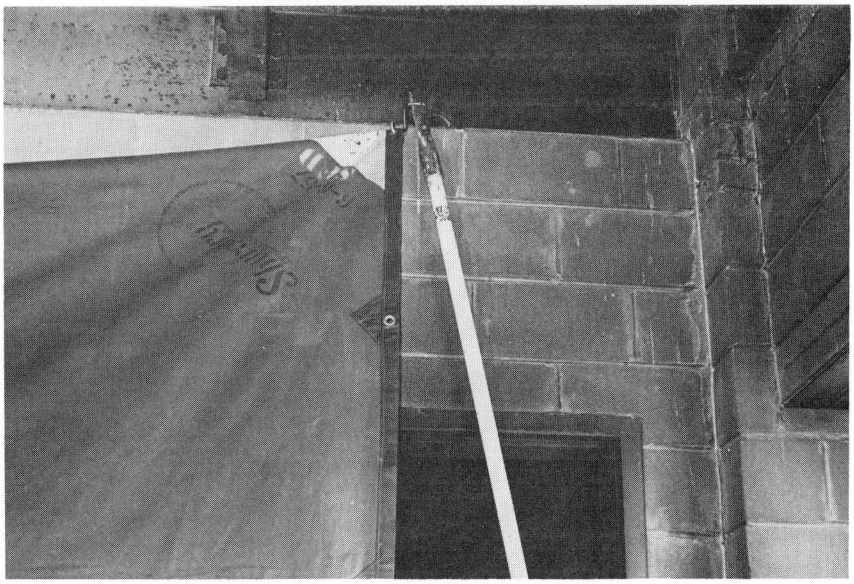

Figure 2-16 In the covering of high shelves or equipment, the Ike-O-Hooks are useful.

3

Methods of Folding and Spreading Salvage Covers

There are several methods of preparing salvage covers so that they can be quickly and easily used at fires. It is strictly a matter of opinion which fold or roll is best adapted to certain conditions. The methods of preparing covers that are included in this manual are the one-person roll, the one-person fold, the two-person fold and the accordion fold. It is is not necessary to develop speed when folding salvage covers, because folding is not an emergency operation. In fact, it is better to take more time and do a neat and careful job.

Instruction and drills are necessary to teach firefighters the art of spreading salvage covers at fires. Teamwork and a knowledge of the principles of salvage practices are also essential. While it is impossible to arrange a drill area that conforms to all conditions likely to be found at fires, it is possible to arrange an area so that the techniques of spreading salvage covers may be practiced. Various methods of spreading covers from each of the folds will be explained and illustrated.

Since a salvage cover cannot effectively be used until furniture or merchandise has been arranged, it is advisable to bend a folded cover in the middle and stand it on the floor at one side of the door until it is needed. If a rolled salvage cover is used, it can also be stood on end and leaned against the wall. Both procedures are shown in Figure 3-1. By standing covers in either manner, only a small portion may be soiled should the floor be wet or dirty and they can be easily found when needed.

NOTE: The one-person or two-person designations refer to how many individuals it will require to unfold or unroll the cover for use. All folds and rolls are usually prepared by two persons.

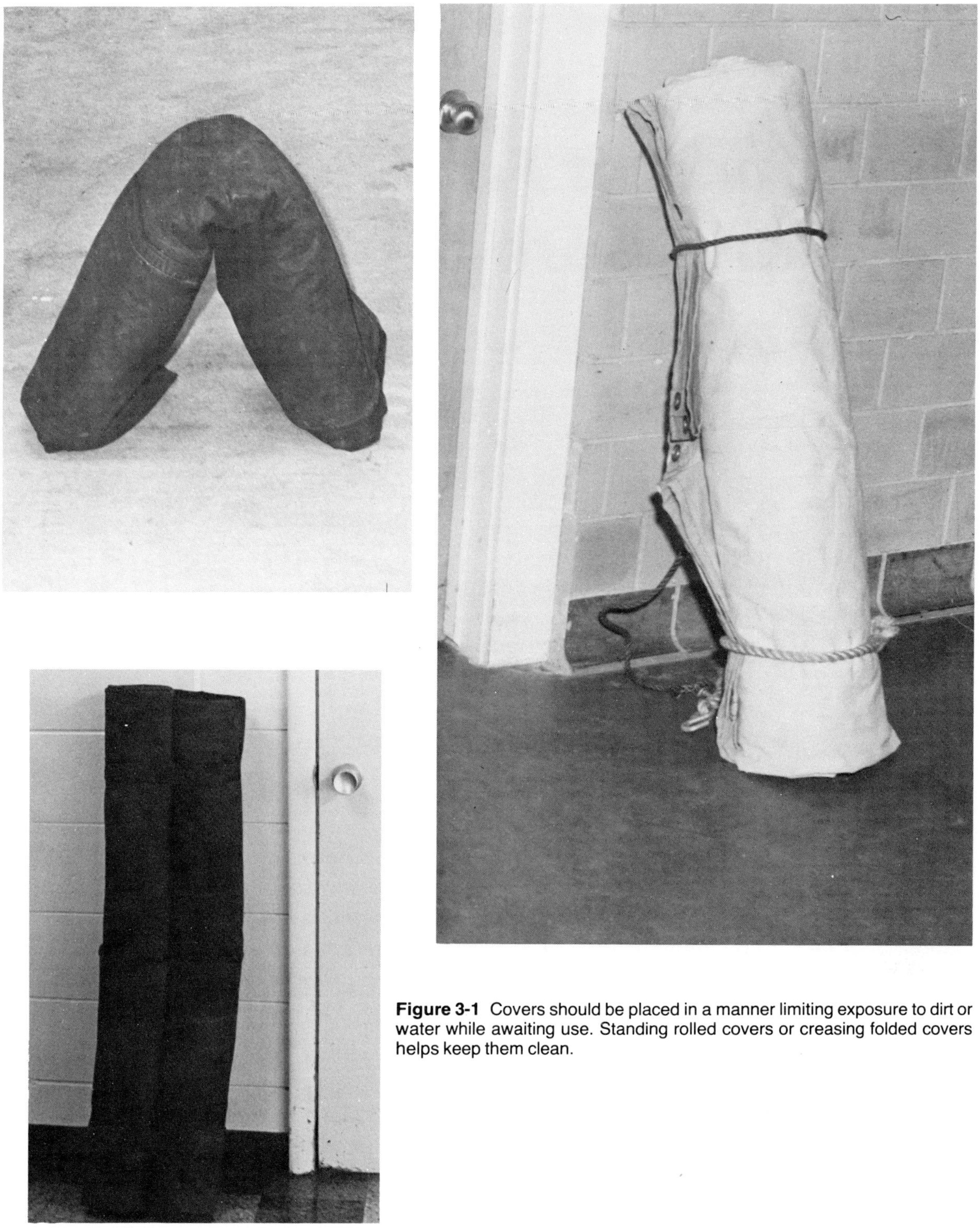

Figure 3-1 Covers should be placed in a manner limiting exposure to dirt or water while awaiting use. Standing rolled covers or creasing folded covers helps keep them clean.

PREPARING THE ONE-PERSON SALVAGE COVER ROLL

The one-person salvage cover roll is used by some departments in preference to the one-person salvage cover fold. Its principal advantage is that it can be rolled across the top of an object and then be opened by one person. To form this roll, the operators must make initial folds to reduce the width of the cover. The following steps for making the one-person roll are illustrated in Figures 3-2 to 3-11.

Step 1: Lay the cover out in a flat position on a clean floor, with the finished (weather) side up. Place an operator at each end to fold one side at a time (Figure 3-2).

Step 2: Grasp the cover with one hand midway between the center and the edge to be folded. Using the other hand as a pivot, pull tightly with the folding hand and bring the fold over toward the center of the cover in a pleating fashion (Figure 3-3).

Figure 3-2 Lay out cover with finished side up.

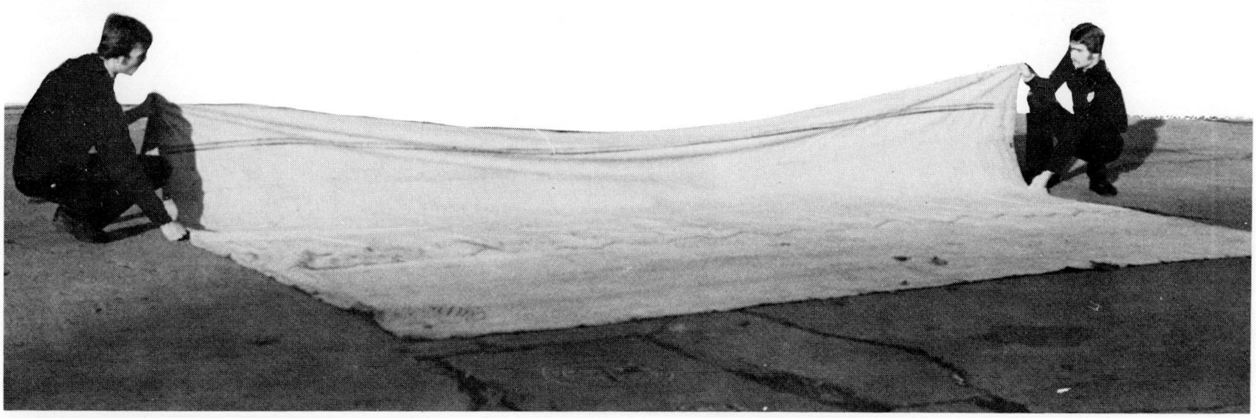

Figure 3-3 Grasp cover midway between the center and edge being folded and form a pleat.

Step 3: After the first fold has been placed at the center, grasp the open edge with the outside hand. Using the other hand as a pivot, stretch the cover tight (Figure 3-4).

Step 4: Bring this outside edge over the center and place it on top and in line with the previously placed first fold (Figure 3-5).

Step 5: Fold the other half of the cover in the same manner, by using Steps 1, 2, 3 and 4. If the folds are not straight, they should be straightened as shown in Figure 3-6. The completed fold is shown in Figure 3-7.

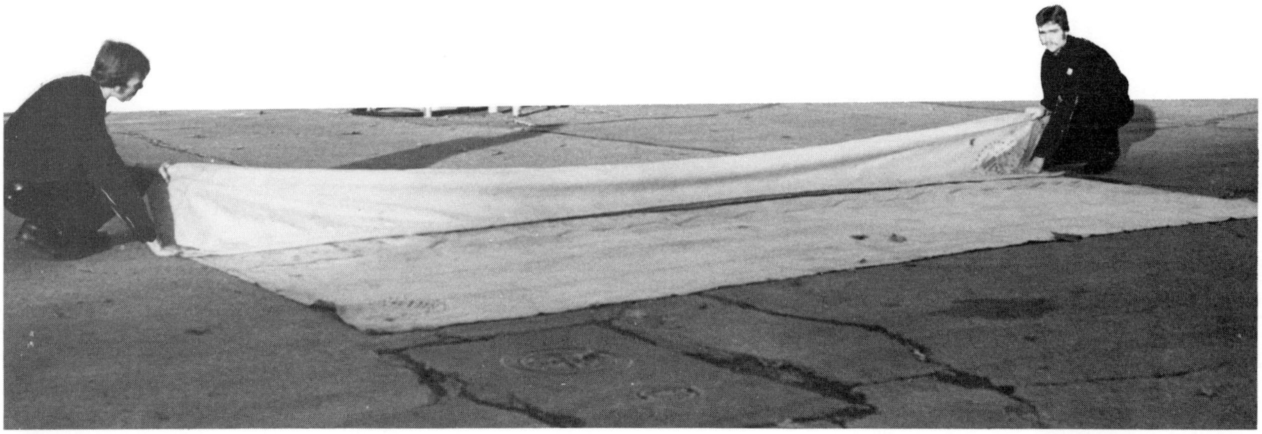

Figure 3-4 Grasp the open edge and stretch the cover tight.

Figure 3-5 Fold the outside edge to the center on top of the first pleat.

Figure 3-6 Repeat previous steps to unfolded side and pull the outside edge to the center on top of the first pleat.

Step 6: Remove air from cover. The folds can be smoothed and air removed by using a yardstick to straighten the folds. A push broom may also be used to remove air from covers (Figure 3-8).

Step 7: Fold over about twelve inches at both ends of the cover to prevent the loose ends from flopping and to make a neater roll.

Step 8: Start the roll at either end and compress the first few rolls as tightly as possible (Figure 3-9).

Figure 3-7 The folded cover before rolling.

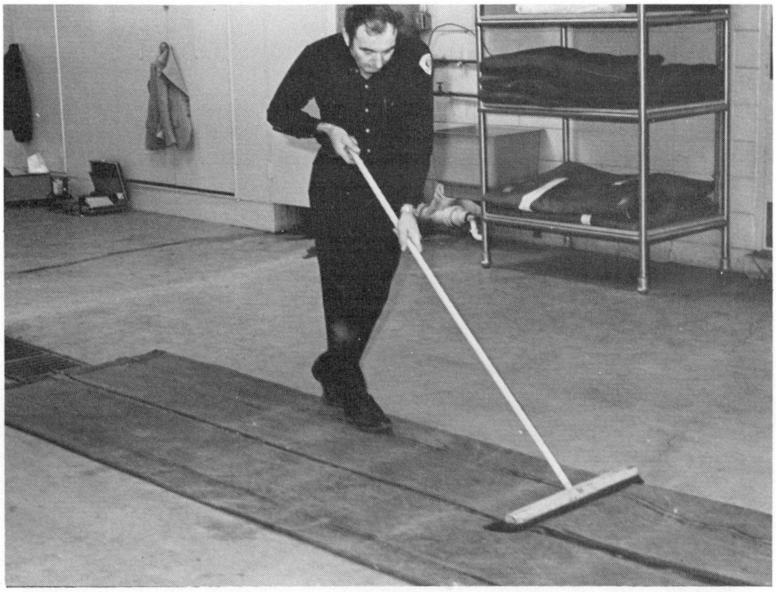

Figure 3-8 A broom can be used to remove air from within the folded cover.

Figure 3-9 Fold cover ends over about 12 inches, and then start rolling.

Step 9: As the roll progresses, tuck in the slack at the center when wrinkles appear (Figure 3-10).

Step 10: The completed roll, as shown in Figure 3-11, may be held secure by rubber bands or tied with cords.

Figure 3-10 Tuck slack in at center when wrinkles appear.

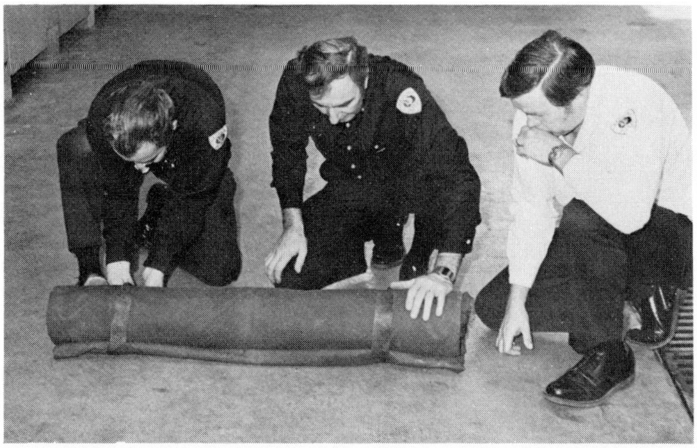

Figure 3-11 Secure the roll with rubber bands or tie with cords.

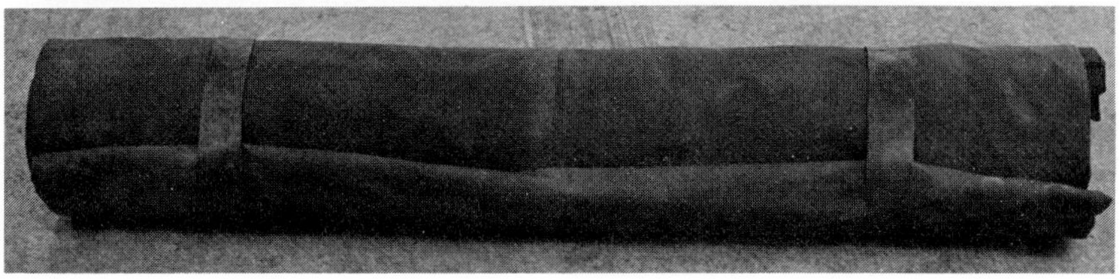

Preparing the Alternate Double-Roll Method

Steps 1-7: Reference: Preparing The One-Person Salvage Cover Roll, Steps 1 through 7.

Step 8: Instead of starting the roll at only one end (Figure 3-9) start the roll from both ends and roll toward the center as shown in Figure 3-12.

Figure 3-12 Start rolling at both ends for the double-roll.

Step 9: As the rolls progress, tuck in any slack at the center.

Step 10: The completed roll, shown in Figure 3-13, may likewise be held in place by rubber bands or cords.

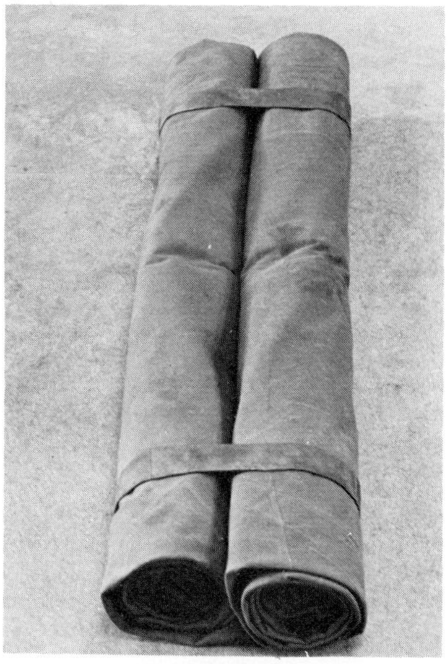

Figure 3-13 The completed double-roll can also be secured with large rubber bands or cords.

SPREADING THE ONE-PERSON ROLL

The one-person salvage cover roll may be carried on the shoulder or under the arm. If it is fastened securely with rubber bands or with cords, its compactness makes it useful in several ways. One use, other than covering objects, is as a floor runner, which can be opened out wide if necessary. The following steps for opening the one-person roll for use are illustrated in Figures 3-14 to 3-19.

Step 1: Start at one end of the object to be covered and while still holding the roll in the hands, unroll it sufficiently to cover the end (Figure 3-14).

Step 2: Lay the roll on the object and continue to unroll toward the opposite end (Figure 3-15).

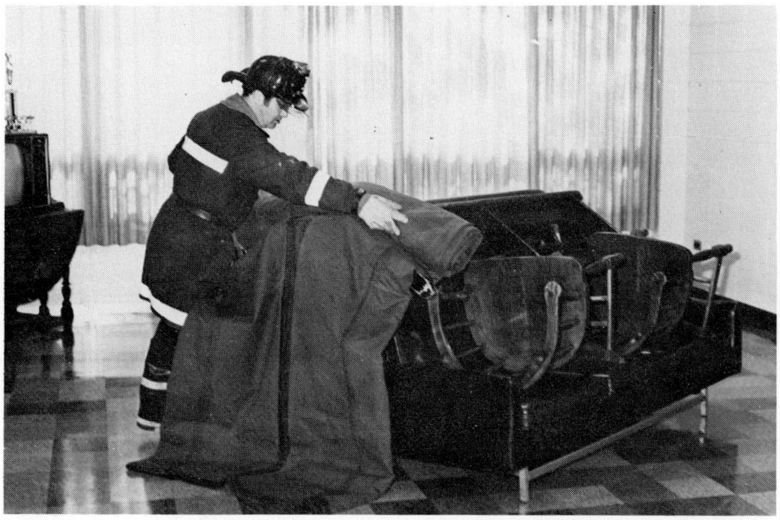

Figure 3-14 Unroll cover to protect end of item being covered.

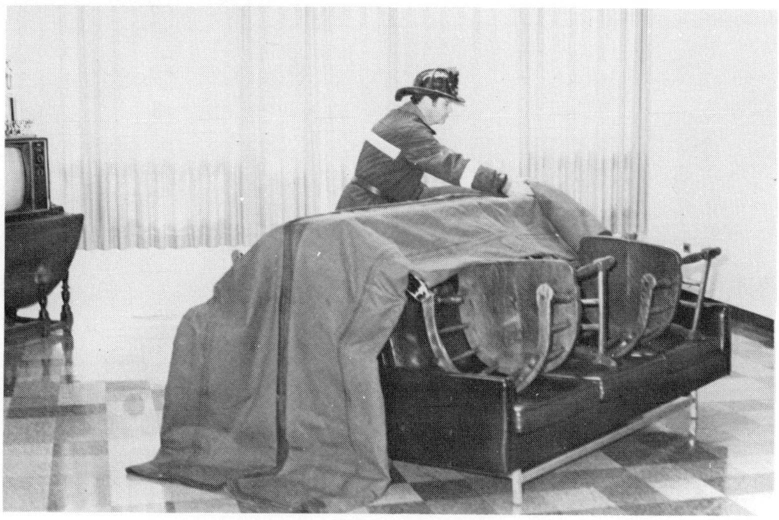

Figure 3-15 Continue to unroll cover over object.

Step 3: Let the rest of the roll either fall into place at the other end or arrange it in position as shown in Figure 3-16.

Step 4: Stand at one end and grasp the open edges where convenient, one edge in each hand (Figure 3-17).

Step 5: Open the sides of the cover over the object by snapping both hands up and out (Figure 3-18).

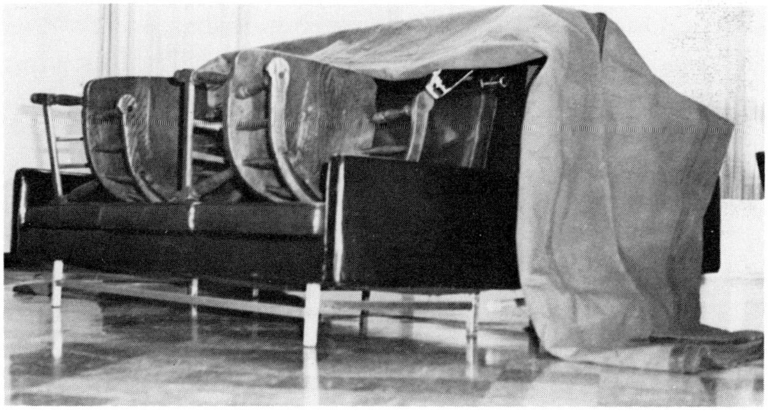

Figure 3-16 Allow roll to fall in place at opposite end.

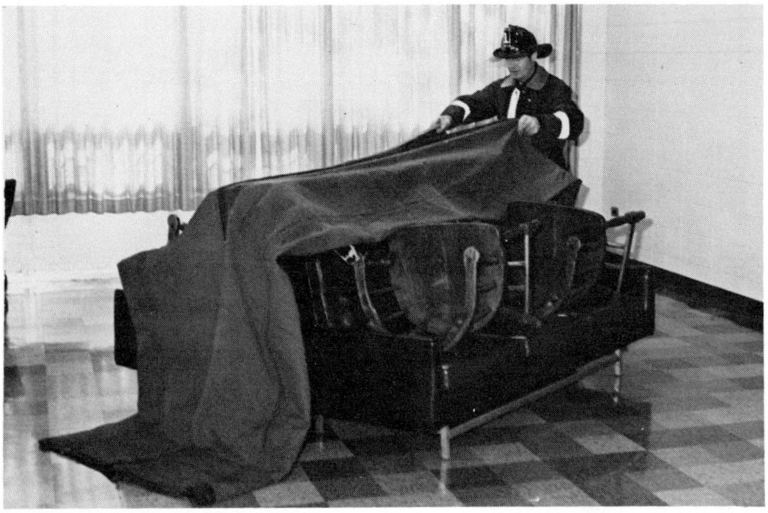

Figure 3-17 Group open edges to unfold cover.

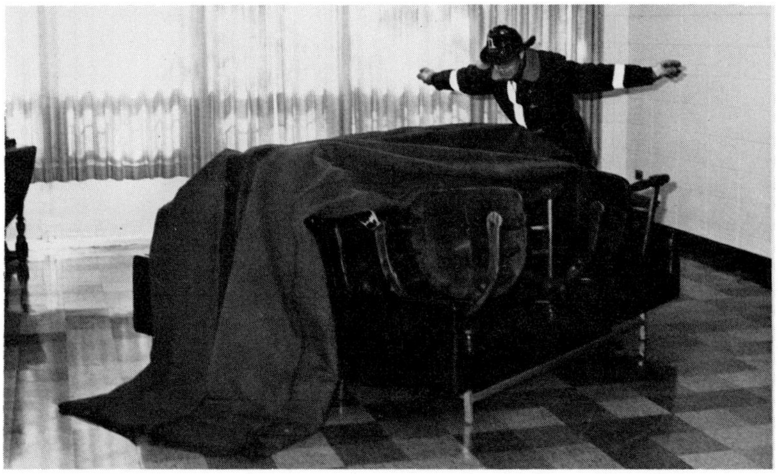

Figure 3-18 Open cover snapping hands up and out.

Step 6: Open the other end of the cover over the object in the same manner (Figure 3-19).

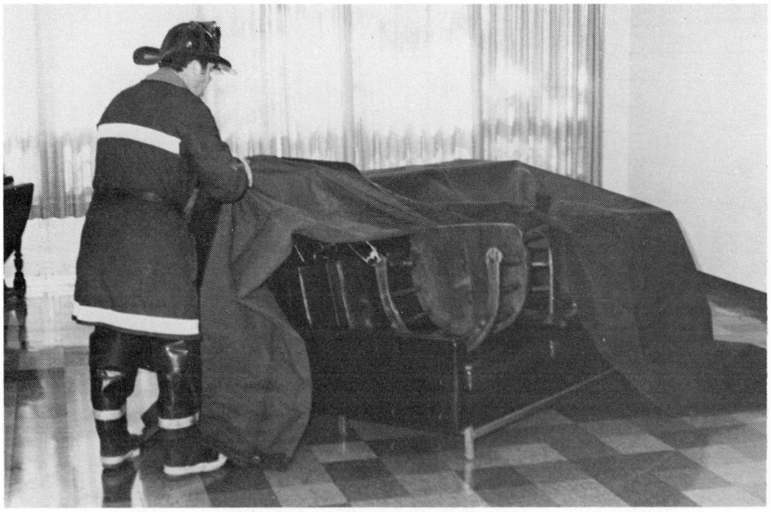

Figure 3-19 Open the other end of the cover in the same manner.

Spreading The Alternate Double-Roll Method

Step 1: Stand at the center of the object to be covered and place the double-roll on top (Figure 3-20).

Step 2: Roll one end of the cover over an end of the object (Figure 3-21).

Figure 3-20 Place double-roll on the center of objects being covered.

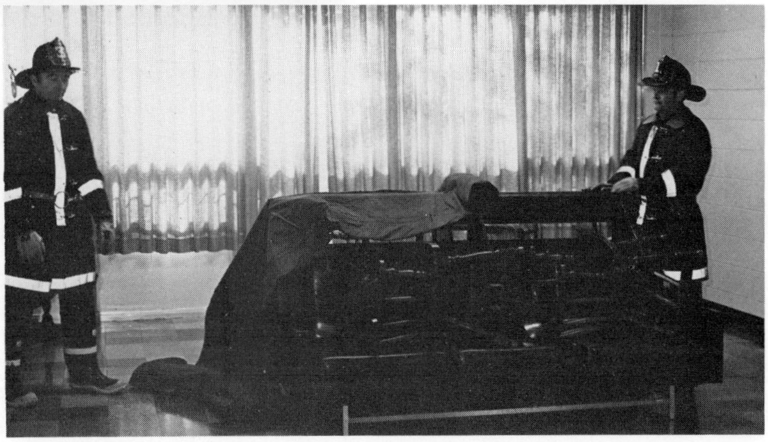

Figure 3-21 Unroll to cover either end.

Step 3: Roll the other end of the cover over the object and arrange both ends in position (Figure 3-22).

Steps 4-6: Reference: Spreading The One-Person Roll, Steps 4 through 6.

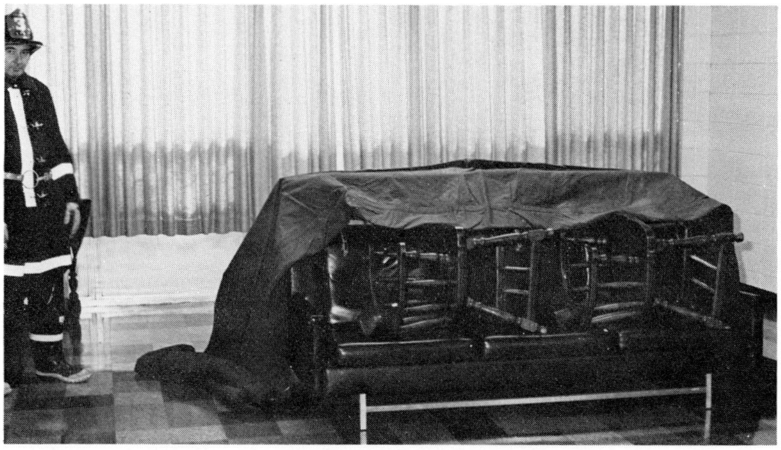

Figure 3-22 Unroll opposite end and prepare to spread cover.

PREPARING THE ONE-PERSON SALVAGE COVER FOLD

The one-person salvage cover fold cannot be opened for use quite as rapidly as a two-person fold, but when personnel is limited it works to a considerable advantage. Like the one-person roll, initial folds must first be made to reduce the width of the cover. These initial folds are exactly the same as those described in Steps 1 through 6 for preparing the one-person roll. The following steps for the one-person salvage cover fold (7 through 11) are illustrated in Figures 3-23 to 3-27.

Figure 3-23 Using the foot as a pivot, bring the folded cover to a point just short of the center.

Steps 1-6: Reference: Preparing The One-Person Salvage Cover Roll, Steps 1 through 6.

Step 7: With the cover folded to a reduced width, both persons grasp one end of the cover and bring it to a point just short of the center. Each person uses his inside foot as a pivot to form the fold (Figure 3-23).

Step 8: Each man then uses one hand as a pivot and brings the folded end over and on top of the previous fold (Figure 3-24).

Step 9: Continue, once again, and bring the folded ends over and on top of the previous fold at the middle (Figure 3-25).

Step 10: Fold the other end of the cover toward the center and leave about four inches between the two folds (Figure 3-26).

Step 11: The space between the folds now serves as a hinge as the operators place one set of folds on top of the other for the completed one-person fold (Figure 3-27).

Figure 3-24 Using the hand as a pivot, bring the folded end to the center.

Figure 3-25 Make a third fold to the center further reducing the size.

Figure 3-26 Fold the other side in a similar manner with four inches between folds.

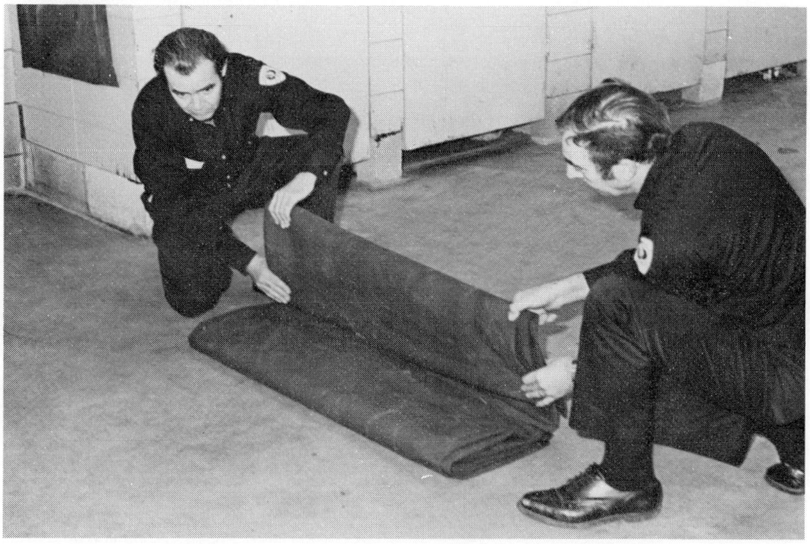

Figure 3-27 Complete the fold by placing one set of folds on top of the other.

SPREADING THE ONE-PERSON FOLD

The one-person salvage cover fold may be carried in any manner that the operator may choose, for he is the only one who will be involved in spreading the cover. It is suggested, however, that this fold be carried on the shoulder for convenience. The following steps for opening the one-person fold for use are illustrated in Figures 3-28 to 3-33.

Step 1: Lay the folded cover on top of and near the center of the object to be covered and separate it at the first fold (Figure 3-28).

Step 2: Select either end and continue to unfold by separating the next fold (Figure 3-29).

Step 3: Continue to unfold this same end toward the end of the object to be covered.

Step 4: Grasp the end of the cover near the center with both hands to prevent the corners from falling outward (Figure 3-30).

Figure 3-28 Place cover at center of object to be covered and separate the sets of folds.

Figure 3-29 Start at either end and cover it by unfolding the cover.

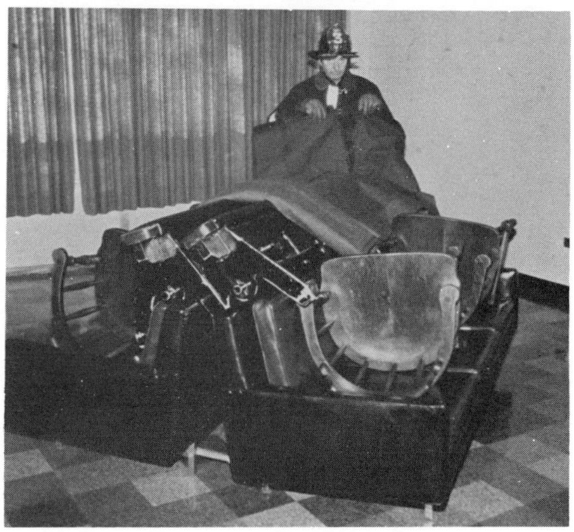

Figure 3-30 Grasp the cover at the center and bring it into position over the end of the object.

Step 5: Bring the end of the cover into position over the end of the object being covered.

Step 6: Unfold the other end of the cover over the object in the same manner (Figure 3-31).

Step 7: Grasp the open edges of the cover at either end and prepare to open the cover as described for the one-person roll (Figure 3-32).

Step 8: Open the sides of the cover over the object by snapping both hands up and out.

Step 9: Open the other end of the cover over the object in the same manner (Figure 3-33).

Two persons can also stretch the cover out and use the normal two-person throws with a cover folded in this manner.

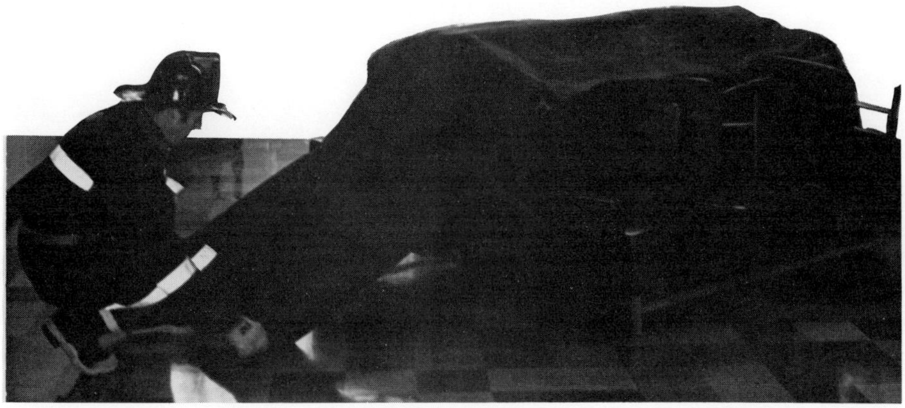

Figure 3-31 Unfold the other end in a similar manner.

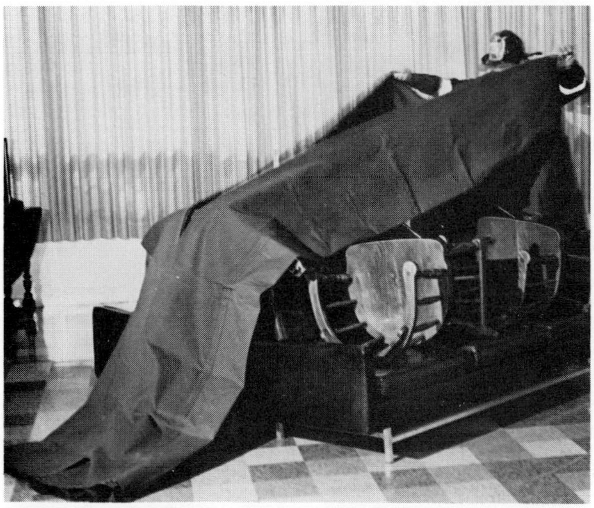

Figure 3-32 Grasp the open edge and spread the cover.

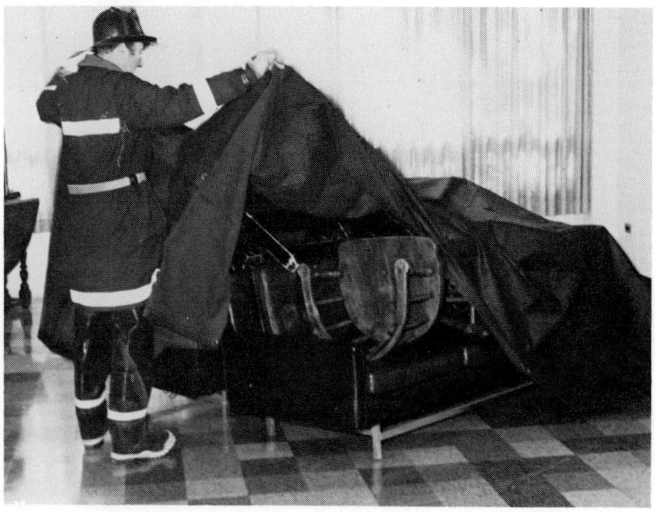

Figure 3-33 Open both ends by snapping the hands up and out to cover the object.

PREPARING THE ACCORDION FOLD

The accordion method of reducing the overall size of the cover is very similar to the method used in the one-person fold, with a slight difference. This difference is that the accordion method folds each half onto itself where the one-person fold tends to pleat each half onto itself. The advantages or disadvantages are a matter of personal preference.

Step 1: Lay the cover in a flat position on the floor with the finished (weather) side up (Figure 3-34).

Step 2: Turn all corners in about ten inches. Then, kneeling as shown in Figure 3-35, place the outside hands on the end of the cover three feet from the edge, reach over with the inside hands and grasp the turned end corners. (The corners are turned in only to provide a handhold; they are flat in the final fold.)

Figure 3-34 Place the cover on the floor with the finished side up and the corners turned in.

Figure 3-35 Use the hand as a pivot, three feet from the cover end, and grasp the turned corner.

Step 3: Working together, bring the outside edge to the center (Figure 3-36).

Step 4: Further reduce the width of the cover by grasping the folded edge, pull taut and bring to the center of the cover (Figure 3-37).

Step 5: Repeat steps 2, 3 and 4 working on the unfolded half of the cover (Figure 3-38).

Figure 3-36 Bring the edge to the cover center.

Figure 3-37 Bring the folded edge to the center to further reduce the width.

Figure 3-38 Fold the other side of the cover in the same manner.

Step 6: Remove excess air in cover with push broom or yardstick (Figure 3-39).

Step 7: At one end of the cover, measure in about nine inches, the span of the outside hand to determine the width of the pleat (Figure 3-40).

Step 8: Place thumbs of outside hands under the cover and, with the inside hands spaced about 18 inches, bring the first pleat even with the end of the cover (Figure 3-41).

Figure 3-39 Remove excess air with a broom or yardstick.

Figure 3-40 Measure in about nine inches for the width of the pleat.

Figure 3-41 Using the hand as a pivot, bring the first pleat even with the end of the cover.

Step 9: Continue making accordion pleats until the cover is completely folded (Figure 3-42).

Step 10: Secure the bundle to protect it while in storage on apparatus (optional). Place rubber bands or cords around the folded cover as in Figure 3-43, or place plywood boards (3/8 inches by 9 inches by 35 inches) on top and bottom under rubber bands made of inner tubes.

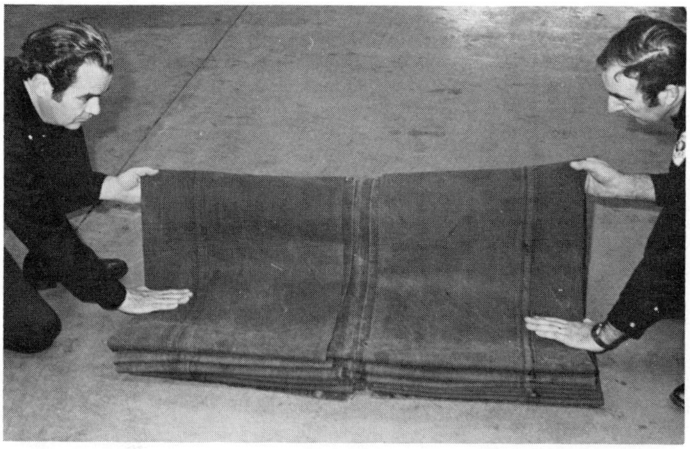

Figure 3-42 Continue making accordion pleats until the cover is folded.

Figure 3-43 The fold can be secured by inner tube bands.

SPREADING THE ACCORDION FOLD WITH ONE-PERSON THROW

The one-person throw should be used only when covering unbreakable materials. It is a rapid method for spreading covers, is particularly advantageous when personnel is limited, and works well when spreading accordion-folded covers. When using the accordion fold, the center of the cover is easily centered on the object. The following steps are utilized for the one-person throw:

Step 1: Center the folded cover on the forearm with the open edges toward the body (Figure 3-44).

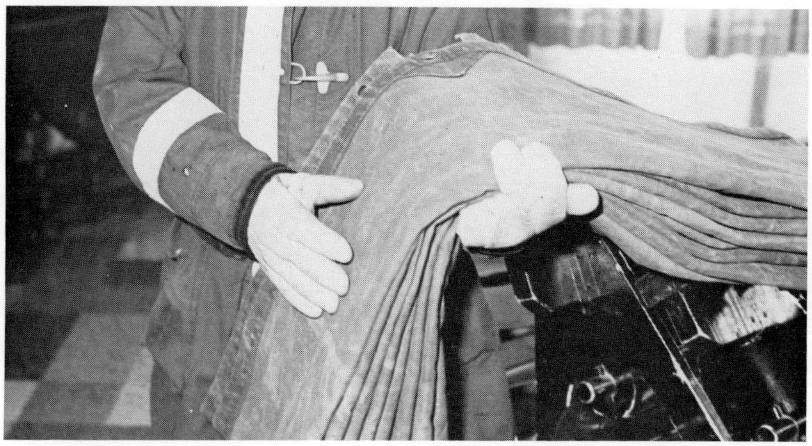

Figure 3-44 Center the cover on the forearm with the open edges toward the body.

Step 2: Lock in the two bottom pleats with the ring and little fingers and hold the index and middle fingers tightly pressed against the remaining pleated edges (Figure 3-45).

Step 3: Grip the three top pleats nearest the body with the thumb and fingers of the throwing hand (Figure 3-46).

Step 4: Hold the cover firmly, and flip the three top pleats over the back of the hand and wrist as the arm is moved into a throwing position. Use the index and middle fingers of the opposite hand as a brake (Figure 3-47).

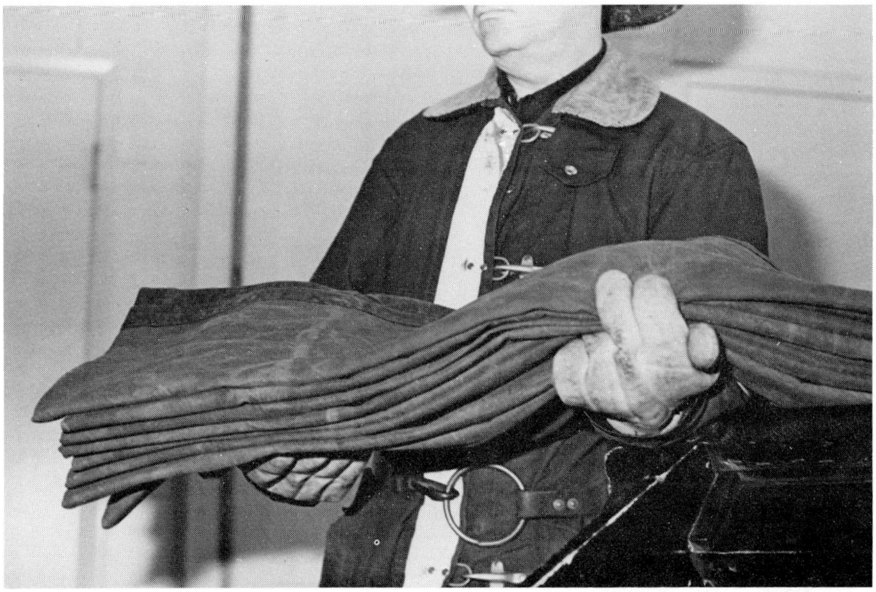

Figure 3-45 Lock the bottom two pleats with the fingers and hold the remaining pleated edges with the middle and index fingers.

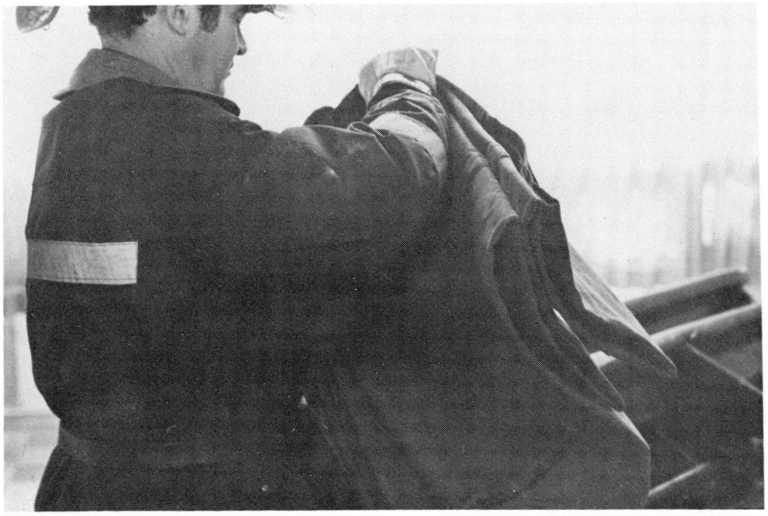

Figure 3-46 Grip the top three pleats with the throwing hand.

Figure 3-47 Place the top three pleats over the hand and wrist as the arm is moved into a throwing position.

Step 5: Throw the top pleats directly overhand. Retain the grip on the bottom pleats, and apply pressure with the index and middle fingers as necessary to control the speed of the throw (Figure 3-48).

Step 6: Open the cover so that it envelops the material, and tuck in all loose edges at the bottom (Figure 3-49).

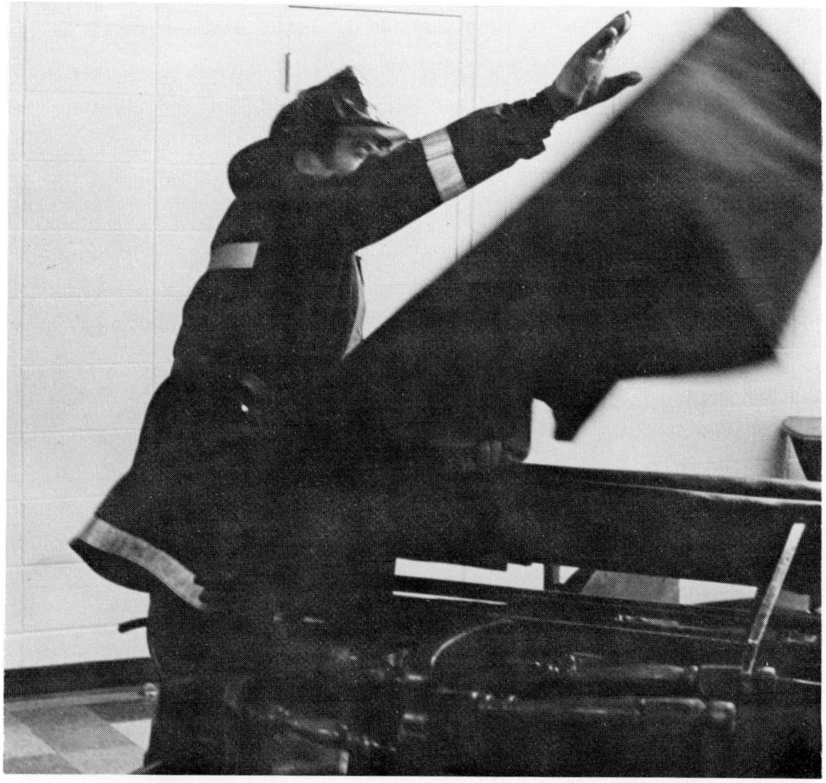

Figure 3-48 Throw the cover in an overhand motion. Use the grip on the bottom pleats to control the throw.

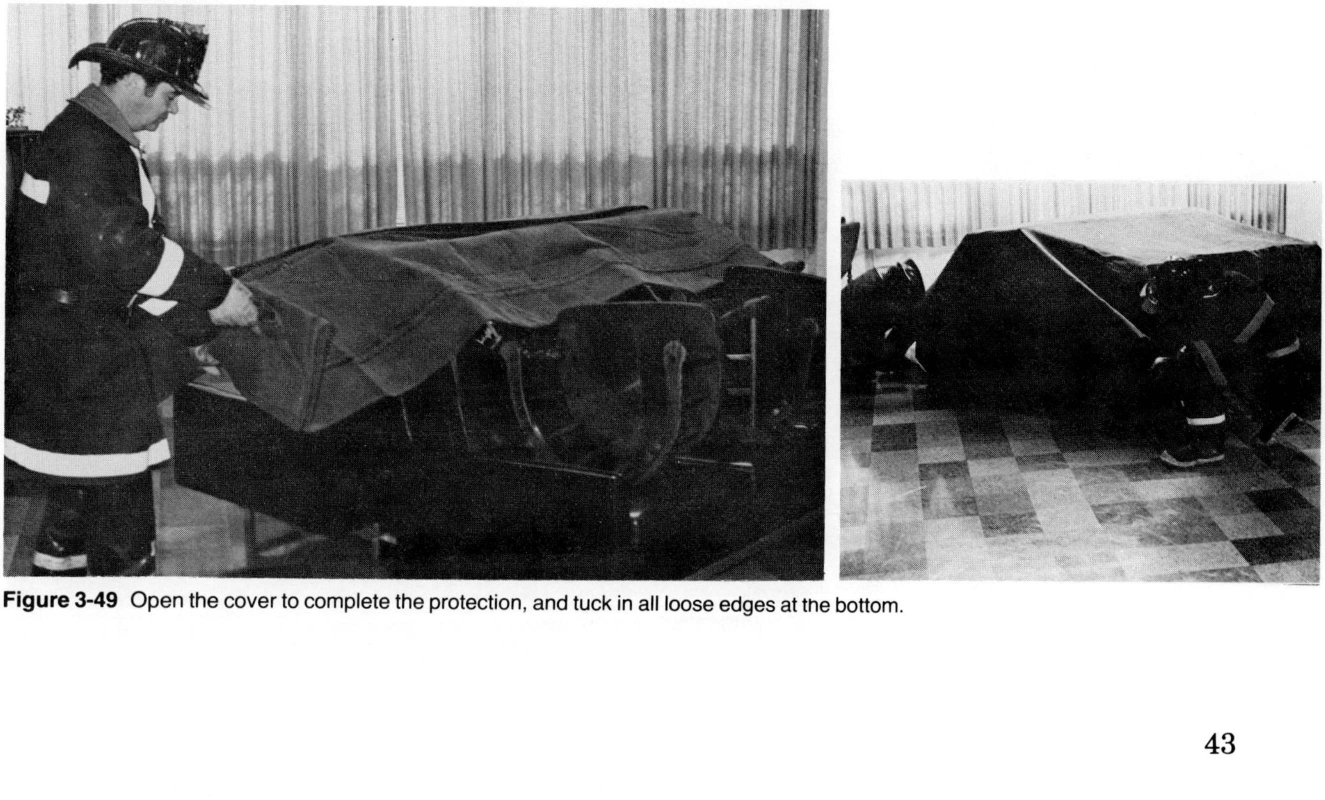

Figure 3-49 Open the cover to complete the protection, and tuck in all loose edges at the bottom.

SPREADING THE ACCORDION FOLD WITH TWO-PERSON COUNTER PAYOFF

The counter payoff, a two-person method of spreading a salvage cover, allows the cover to be placed exactly where it is needed on counters or tables. The cover may also be placed gently on fragile articles. The procedures are as follows:

Step 1: Hold the cover on the forearms, with the fingers gripping the two bottom pleats, while the second person grasps the top layer with both hands (Figure 3-50).

Step 2: Extend the cover and hold the arms high as the second person walks backward (Figure 3-51).

Figure 3-50 Hold the cover by the bottom two pleats while the second person takes the top layer with both hands.

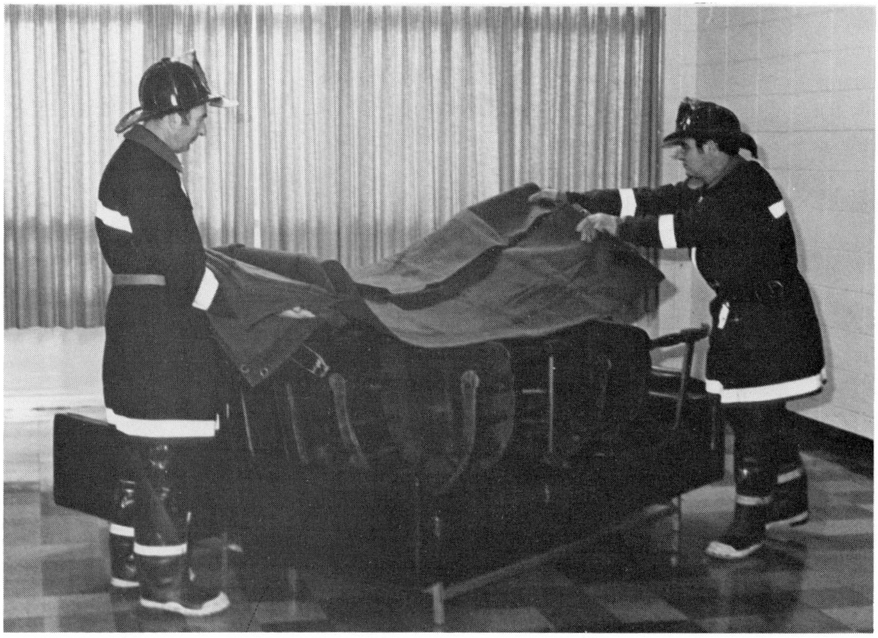

Figure 3-51 The second person extends the cover holding his arms high as he backs along the furnishings.

Step 3: Position the extended cover gently over the object (Figure 3-52).

Step 4: Open the cover and tuck in the edges around the base (Figure 3-53).

Figure 3-54 illustrates the three one-person methods of folding covers described above, the one-person fold, the one-person roll (double) and the accordion fold.

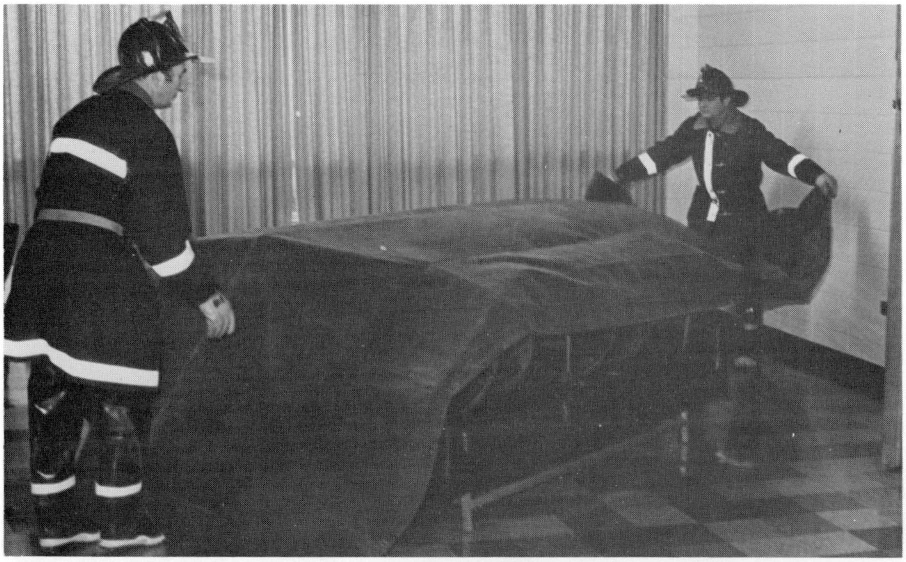

Figure 3-52 Spread the cover gently over the object.

Figure 3-53 Complete the covering and secure corners and edges.

Figure 3-54 Various one-person methods of folding covers.

PREPARING THE TWO-PERSON SALVAGE COVER FOLD

The versatile two-person or "Underwriters" salvage cover fold is widely used because of the speed of application. It is adaptable to several different throws for covering materials and works well for developing catchalls and chutes. The following steps for making the fold are illustrated in Figures 3-55 to 3-66.

Step 1: With the cover stretched lengthwise and clean side down, each person grasps the end of the cover at its center grommet and then pulls the cover tightly between them. Raise this center fold high above the floor and shake out the wrinkles to form the first half fold (Figure 3-55).

Step 2: By holding the cover tight between the two operators it may be swung from side to side and carefully spread upon the floor. If any wrinkles occur they should be removed (Figure 3-56).

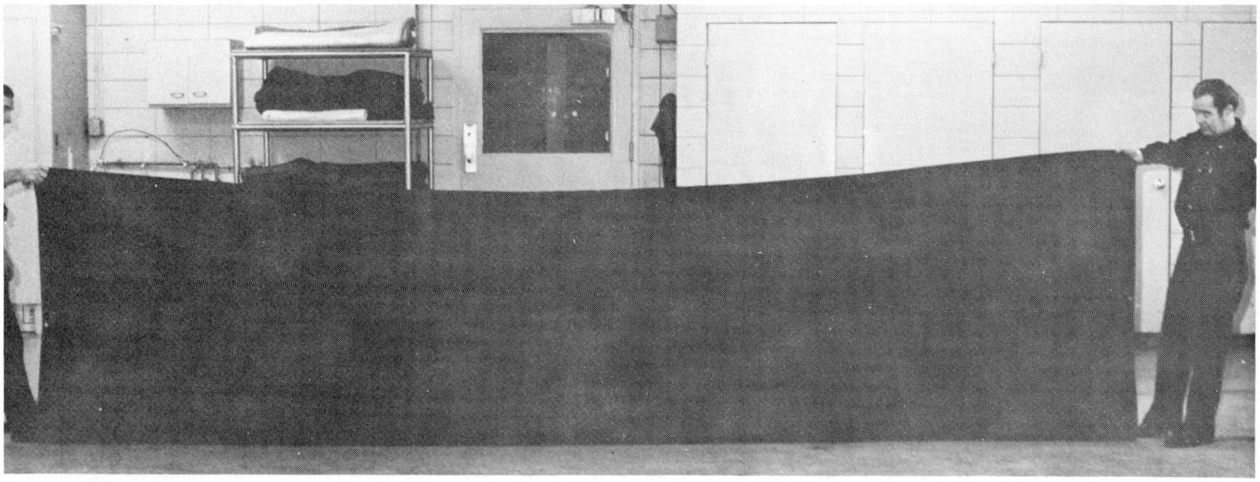

Figure 3-55 Form the first half fold by raising the cover high at the center grommet with finished side to the inside.

Figure 3-56 The cover is spread on the floor and all wrinkles removed.

Step 3: While still grasping the cover near its center grommet, place inside foot at the center of the half-fold and thus make a pivot for the next fold (Figure 3-57).

Step 4: Stretch that part of the cover being folded tightly between the operators. Make the quarter-fold by folding the open edges over the folded edges (Figure 3-58). Alternate method, fold closed edge to open.

Step 5: One person should secure one end of the quarter-fold while the other person grasps the opposite end and shakes out all the wrinkles (Figure 3-59).

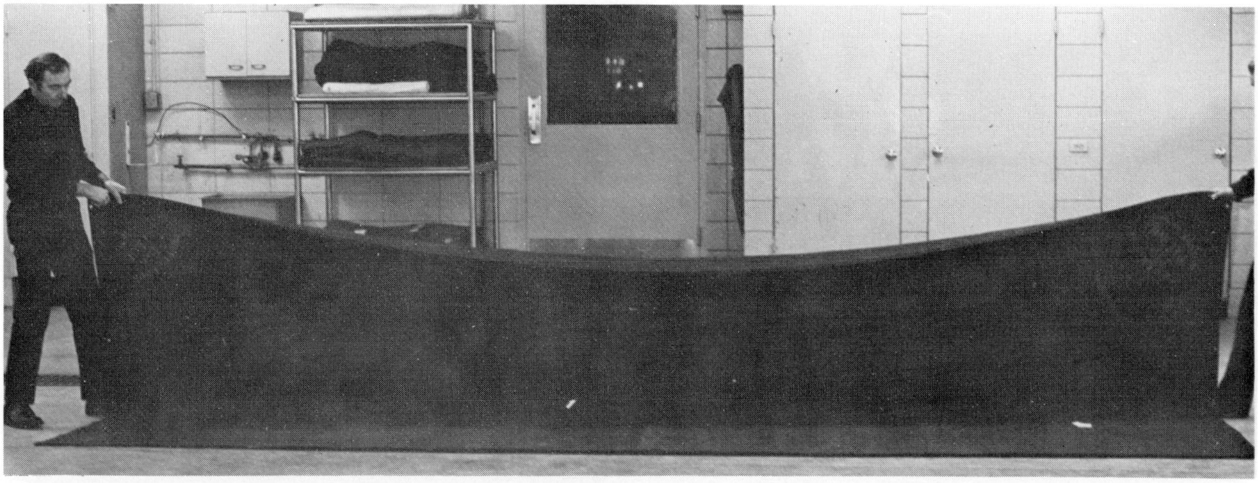

Figure 3-57 Using the feet as pivots at the center of the half fold, prepare to make the quarter-fold keeping the cover taut.

Figure 3-58 Complete the quarter-fold by folding the open edges over the folded edge. By folding the folded edge over the open edges the four corners will be separated by two double folds when the fold is completed.

Figure 3-59 Grasp both ends of the quarter-folded cover and shake out the wrinkles.

Step 6: The person holding the end of the cover snaps the bottom portion forward and carries the quarter-fold to his partner. If the partner is kneeling, he can direct the quarter-fold into position (Figure 3-60).

Step 7: Place one end on top of the other. With a person at each end, the cover can be lifted and pulled taut to remove any wrinkles or gathering (Figure 3-61).

Figure 3-60 Snap the bottom of the cover forward and move to fold it reducing the length.

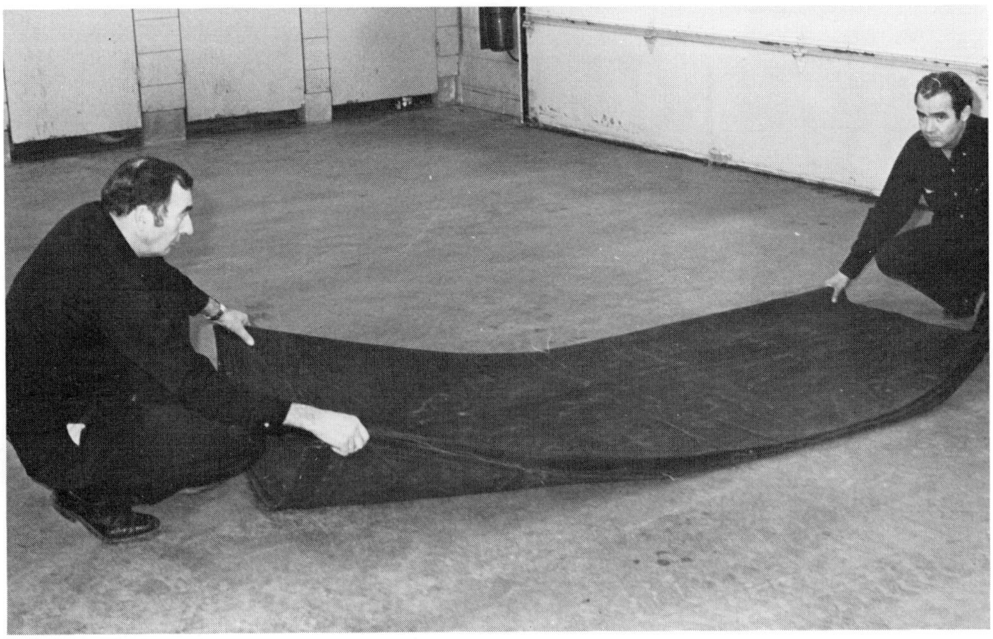

Figure 3-61 Place one end on top of the other and remove any wrinkles.

Step 8: The two operators then grasp the open ends and use their inside foot as a pivot for the next fold (Figure 3-62).

Step 9: Bring these open ends over and place them on top of the folded center fold (Figure 3-63).

Step 10: Continue this folding process by bringing the open ends over to the folded end. During this fold the free hand may be used as a pivot to hold the cover straight (Figure 3-64).

Figure 3-62 The open ends are then folded again using the feet as pivots.

Figure 3-63 Line up the edges and remove any wrinkles.

Figure 3-64 Using the hand as a pivot, continue to fold the open edges over the folded edge.

Step 11: Complete the operation by one more fold in the same manner. Bring the open ends over and to the folded end using the free hand as a pivot during the fold (Figure 3-65).

Step 12: The completed two-person salvage cover fold is shown in Figure 3-66.

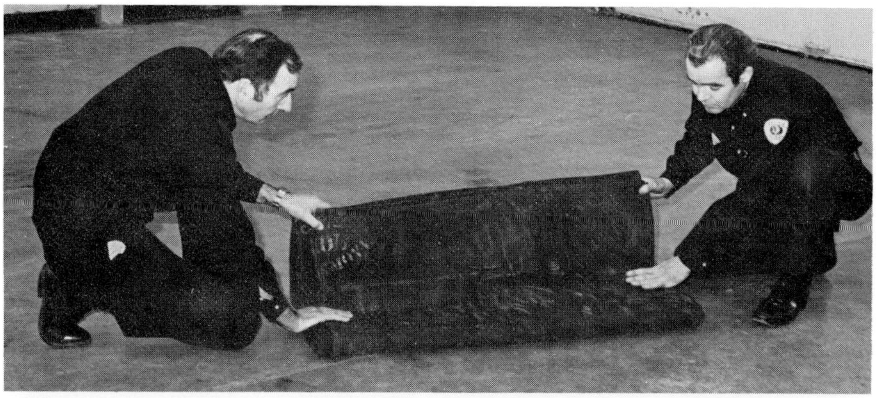

Figure 3-65 One more fold completes the operation with the open edges appearing on top.

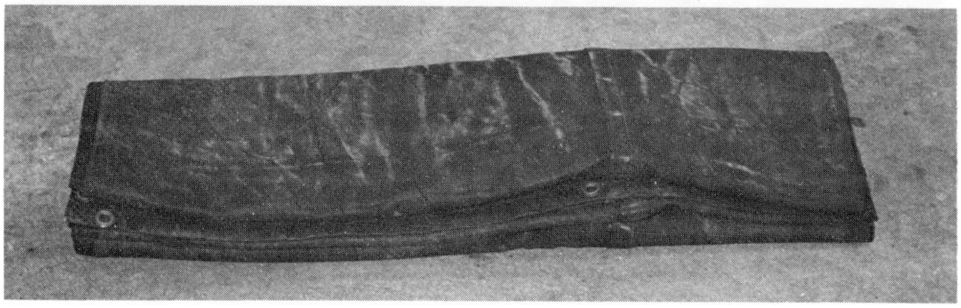

Figure 3-66 The completed two-person salvage cover fold.

CARRYING AND POSITIONING THE TWO-PERSON FOLD

It is important to consider how the two-person salvage cover fold may be carried. Probably the most convenient way to carry this fold is on the shoulder. Covers can be carried on either shoulder with the open side towards the neck and with the single corners or edges toward the front. In this manner it is easy for the carrier to separate the folds and give the top two singles to his partner and grasp the bottom two singles himself. This situation is illustrated in Figure 3-67. It may be preferable to standardize carrying the cover on the left shoulder within a department.

If the cover is placed on the shoulder with the seam in front and the four corners to the rear, the situation as shown in Figure 3-68, will exist. The cover does not need to be turned around to secure the corners at this point. The carrier may separate the two open-end folds with the uppermost fold for his partner and the lower fold for himself as illustrated in Figure 3-69. After the cover has been stretched lengthwise, the corners may be secured by running the hand down the end hem. With a little practice and training, both of these methods of opening a cover can be easily mastered, even in darkness.

Small ropes can be placed in the corner grommets or the corners appropriately painted to assist in identification (Figure 3-70).

Normally, when carrying covers from apparatus to an area where they need to be spread, three covers make a load. But if covers need to be carried any distance at all, it is better to carry only two covers at a time.

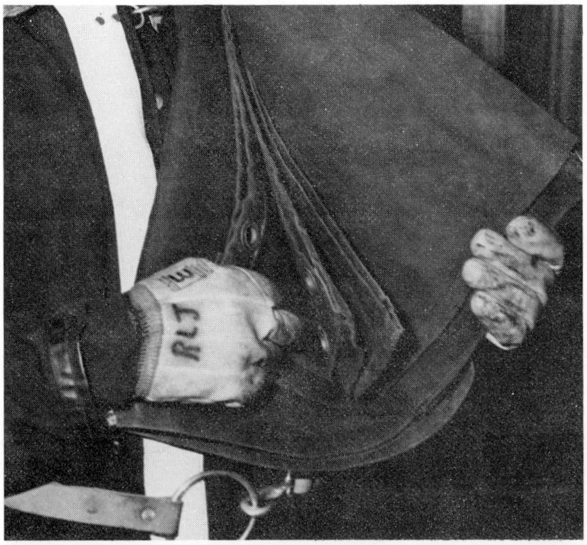

Figure 3-67 The two-person fold is carried on the shoulder so the top two single edges are available to the partner and the bottom two single edges to the carrier. Alternate method of folding separates the top and bottom edges by two double folds.

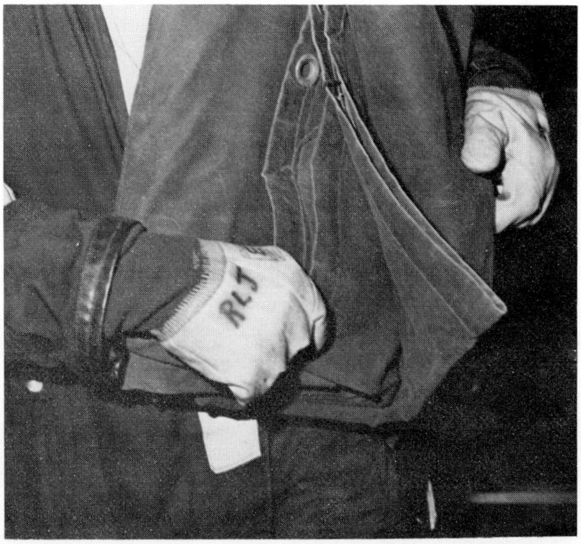

Figure 3-68 Cover carried with corners placed to the rear.

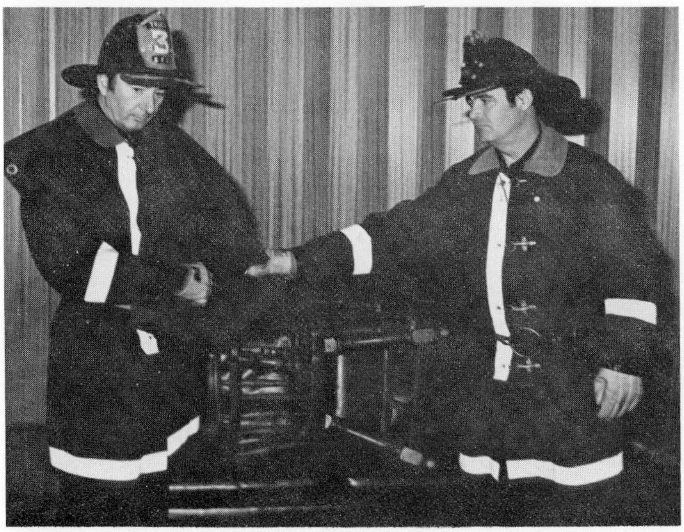

Figure 3-69 With the folds to the front, the top open-end fold goes to the partner and the carrier secures the lower fold.

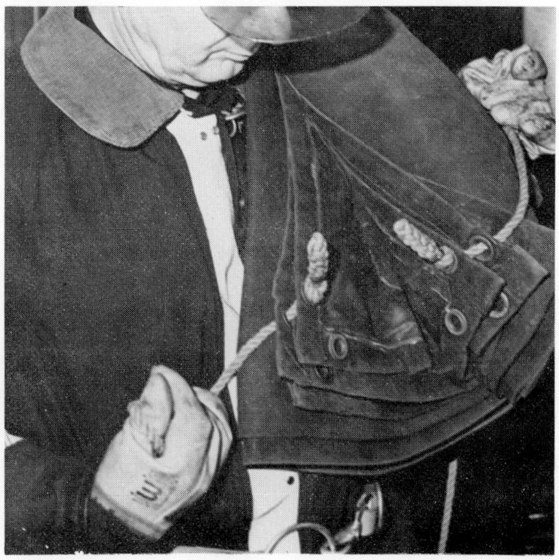

Figure 3-70 Small ropes can be placed in the corner grommets to assist in identification.

SPREADING COVERS WITH THE BALLOON THROW

The balloon throw gives best results when sufficient air is pocketed under the cover. This pocketed air gives the cover a parachute effect which tends to float it in place over the article to be covered. The following steps for making the balloon throw are illustrated in Figures 3-71 to 3-75.

Step 1: Stretch the cover along one side of the object to be covered and separate the last half-fold by grasping each side of the cover near the ends (Figure 3-71).

Step 2: Make several accordion folds in the hand that is to make the throw and place the other hand about midway down the end hem (Figure 3-72).

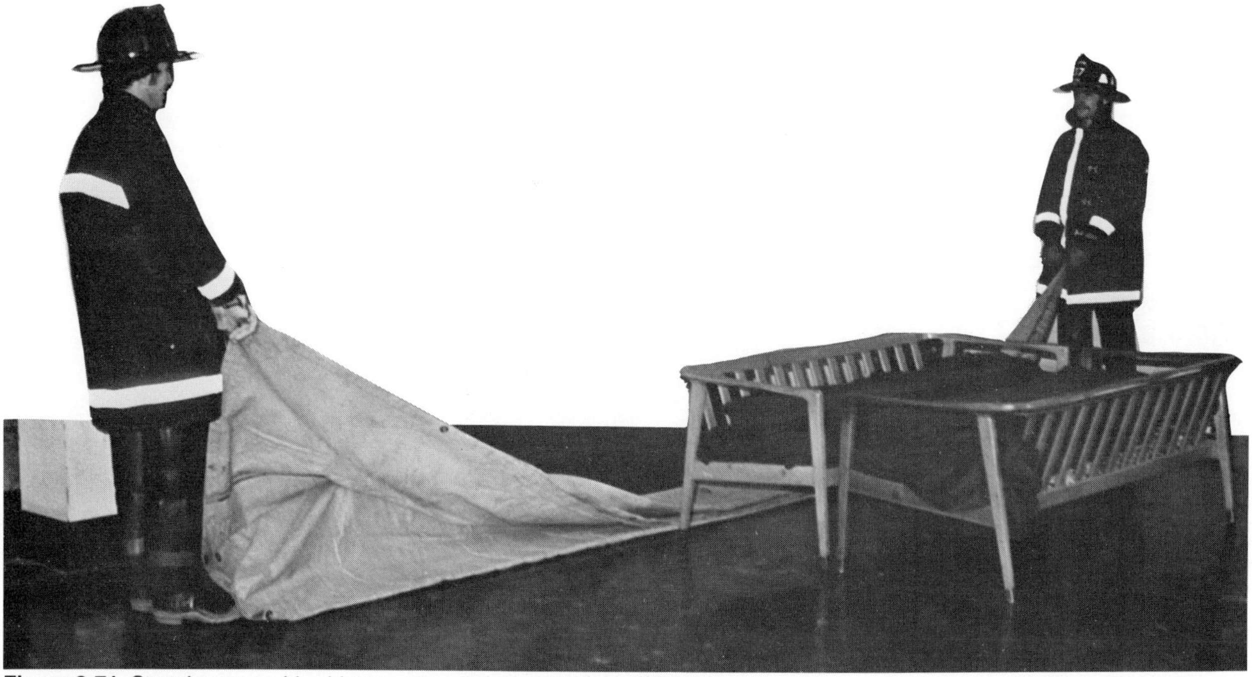

Figure 3-71 Stretch cover aside object to cover and separate the half-fold.

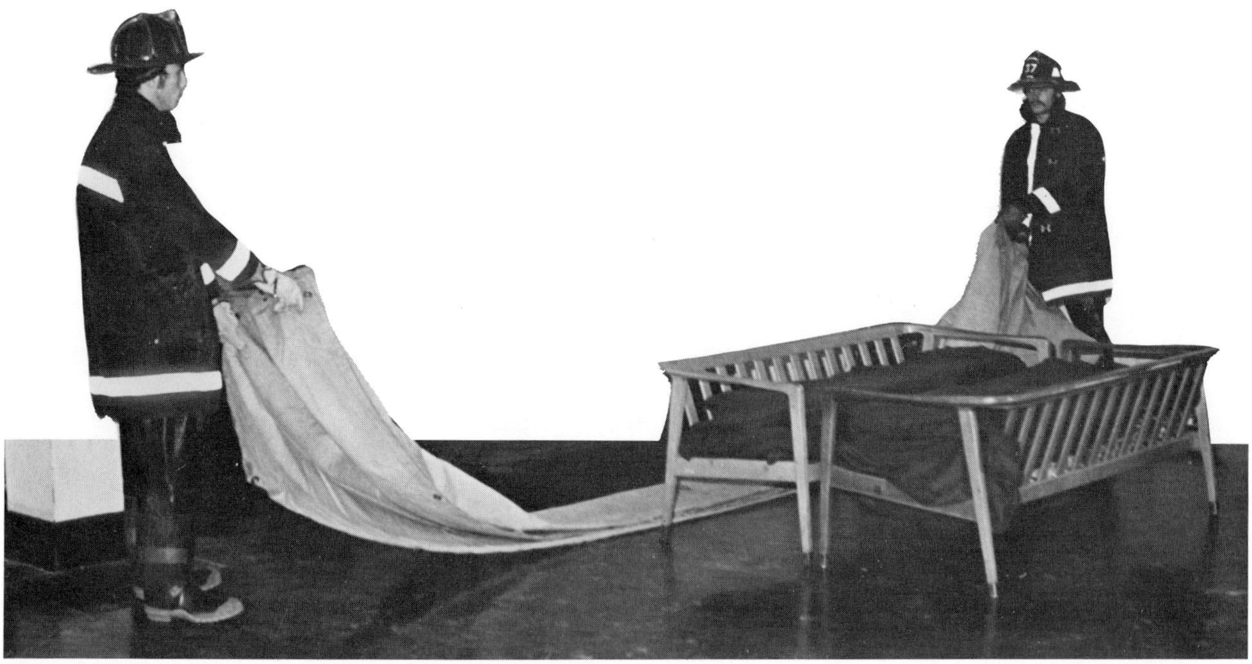

Figure 3-72 Make several accordion folds in the throwing hand.

Step 3: Pull the cover tightly between the operators and prepare to swing the folded part down, out, and then up in one sweeping movement so as to pocket as much air as possible (Figure 3-73).

Step 4: When the cover is as high as the operators can reach, the accordion folds may either be pitched or carried across the object, an action which causes the cover to float over the object (Figure 3-74).

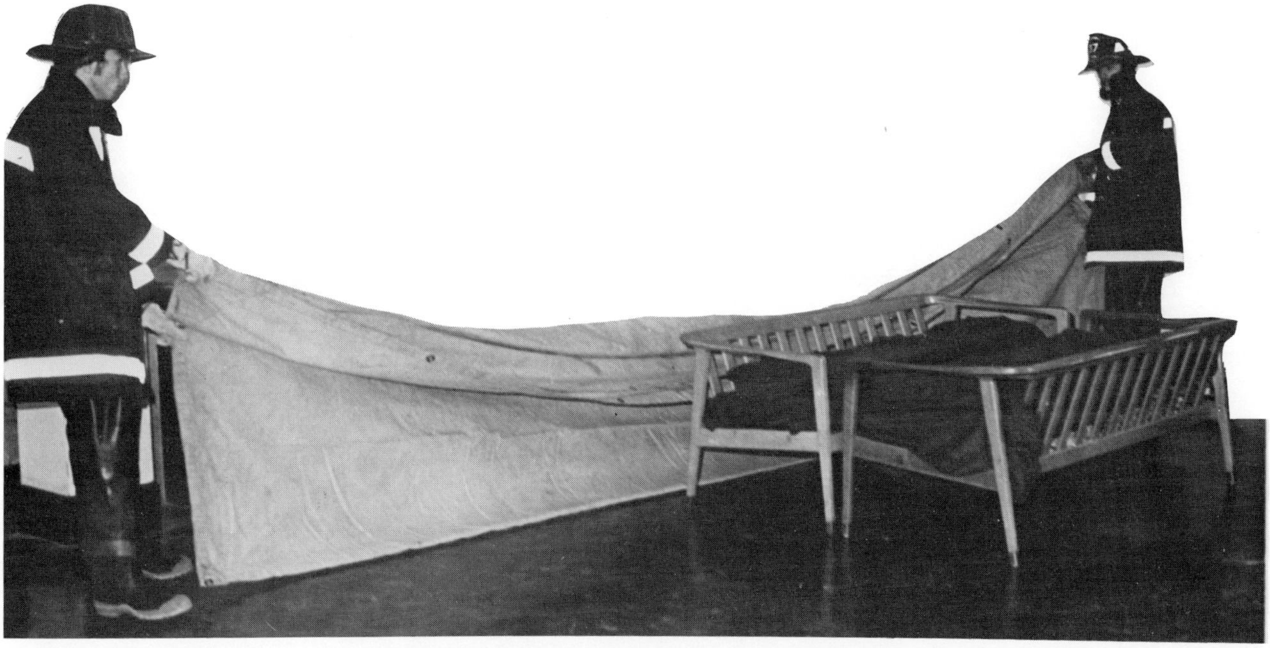

Figure 3-73 Pull the cover taut, and swing down, out and up in a sweeping movement trapping air.

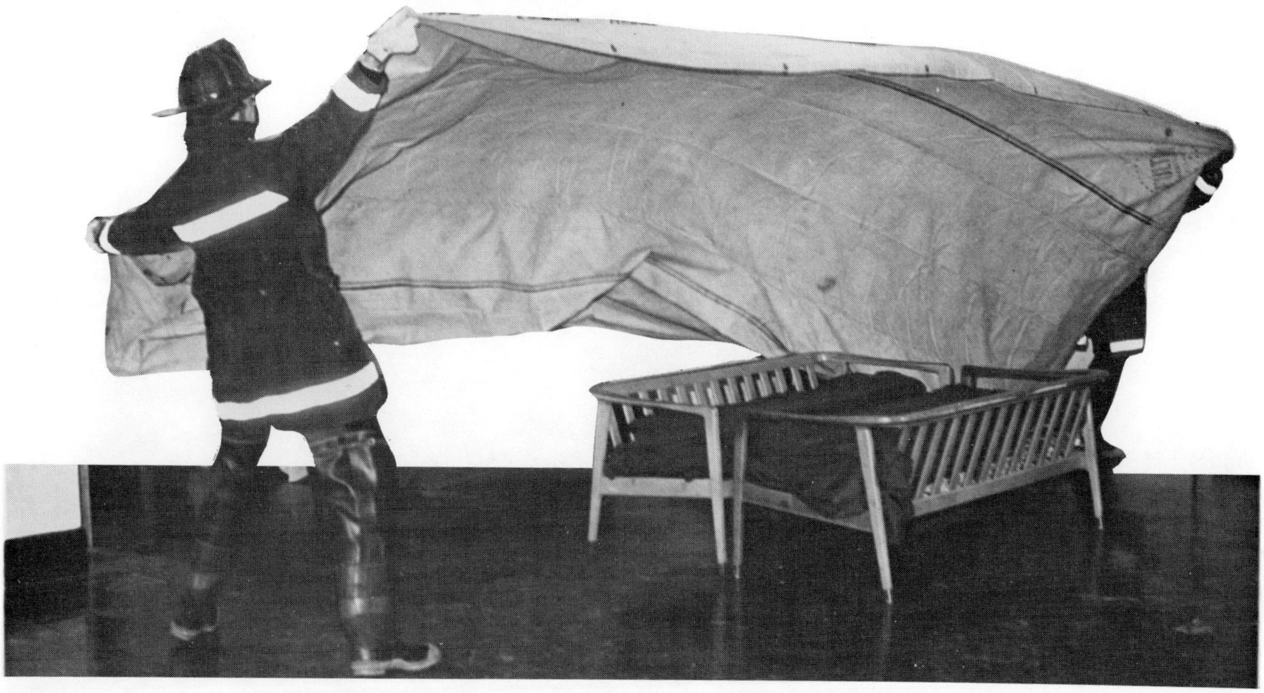

Figure 3-74 With the cover at its peak, the accordion folds are carried across the object.

Step 5: As the cover is floated over the object, guide it into position and straighten for better water runoff (Figure 3-75).

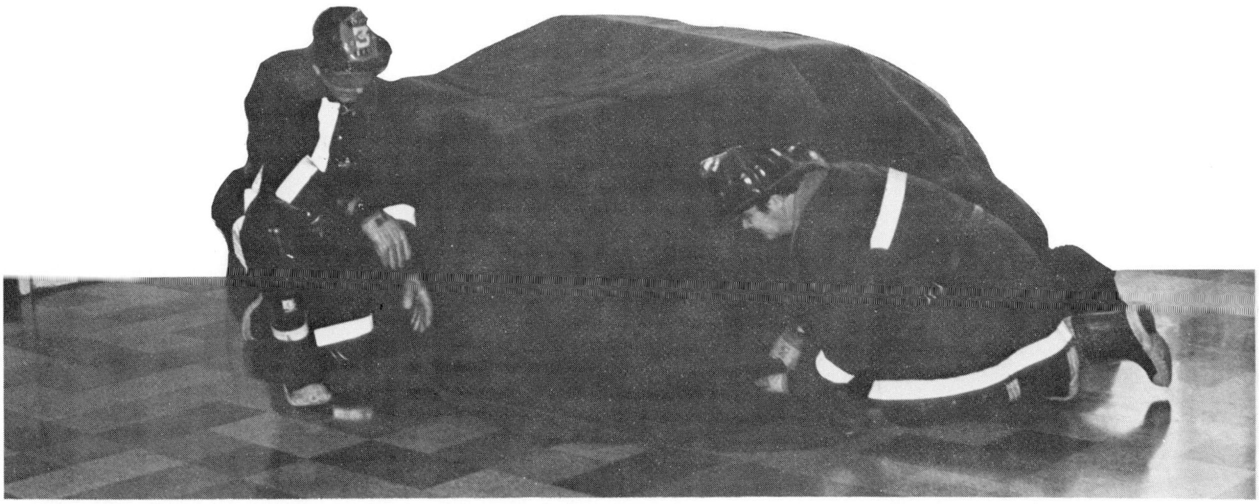

Figure 3-75 Guide the cover into position and straighten for water runoff.

SPREADING COVERS WITH THE SINGLE-EDGE SNAP THROW

The single-edge snap throw can be thrown high and over the object by a snap action. This throw is designed to be used in narrow spaces where the balloon throw cannot be used. It is especially effective where counters or other objects are near a wall or stacked closely together. The following steps are illustrated in Figures 3-76 to 3-80.

Step 1: Stretch the cover along one side of the object to be covered and drop the edge which is next to the object (Figure 3-76).

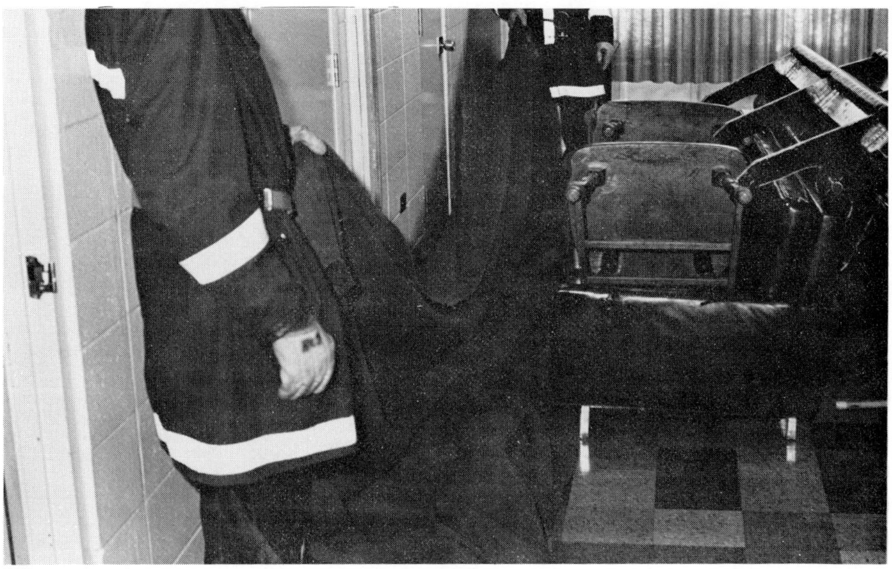

Figure 3-76 Stretch the cover along the furnishings and drop the edge next to them.

Step 2: Bunch the other edge into both hands of each operator so as to form a bulk to be thrown. Pull the cover tightly between the operators and dip the folded part toward the floor ready for the upward thrust (Figure 3-77).

Step 3: Continue in one movement by snapping the cover upward as high as the operators can reach (Figure 3-78).

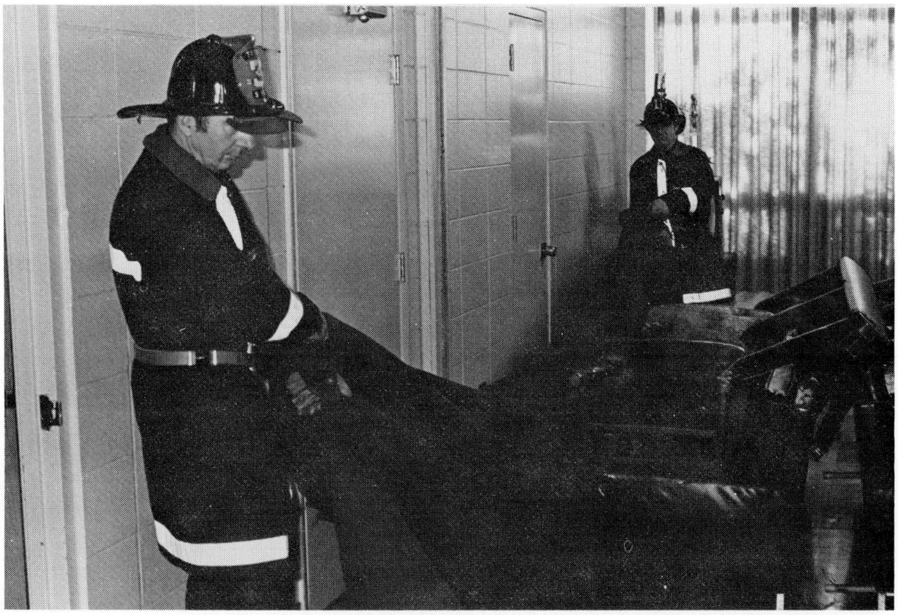

Figure 3-77 Bunch the other edge in both hands, pull tightly and dip the cover ready for the throw.

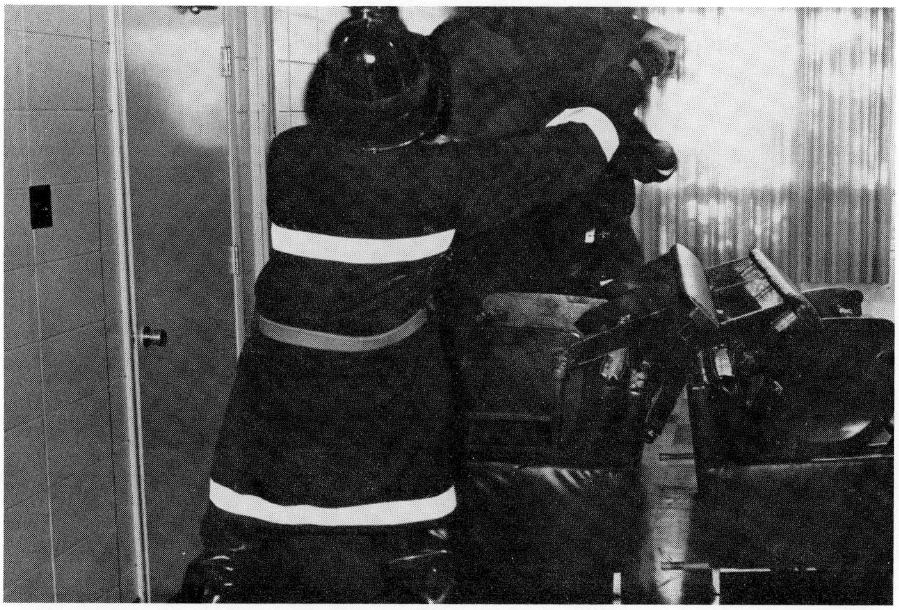

Figure 3-78 Bring the cover up as high as possible.

Step 4: When the cover is higher than the objects to be covered, each operator snaps the folded part from his hand across the object (Figure 3-79).

Step 5: When the cover is well over the object, guide it into position and straighten for better water runoff (Figure 3-80).

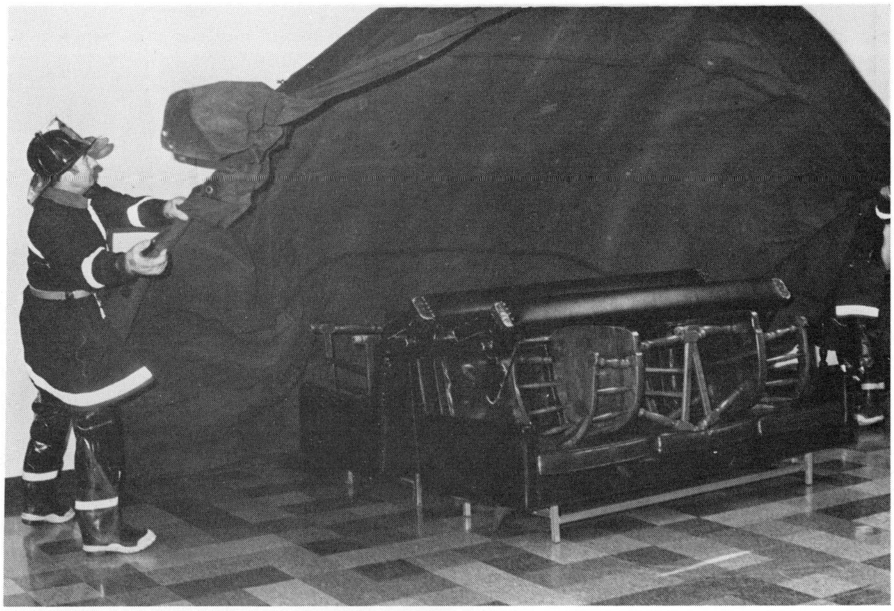

Figure 3-79 Snap the folded part from the hands across the object.

Figure 3-80 Guide the cover into position and straighten.

SPREADING COVERS WITH THE DOUBLE-EDGE SNAP THROW

The double-edge snap throw is limited to special cases where two relatively narrow objects exist on each side of the operators. Depending upon how the cover used is folded, it may be necessary to turn the cover before throwing in order to cover with the clean side. The preparation for the throw consists of bunching the edge of an opened salvage cover into the hands of each operator. This bunching process makes it possible for the bulk of both edges to be thrust upward and then thrown outward by a snap action. Its use between two objects such as counters permits both objects to be covered by one cover or the cover thrown to both sides. The following steps are illustrated in Figures 3-81 to 3-85.

Step 1: Remove the folded cover from the carrier's shoulder as previously described and stretch it between the objects to be covered (Figure 3-81).

Step 2: At a convenient point on the cover edges, bunch or fold both edges of the cover into the hands of each operator leaving approximately three or four feet of the cover between the hands. The cover, in this position, does not touch the floor (Figure 3-82).

Step 3: Pull the cover tightly between the operators, bring the hands close together and with a semi-squatting movement, dip the cover toward the floor ready for the upward thrust (Figure 3-83).

Figure 3-81 Stretch cover between objects which will be covered.

Figure 3-82 Gather both edges of the cover in the operators' hands.

Figure 3-83 Pull the cover tight and dip it in preparation for the upward thrust.

Step 4: Continue in one movement by snapping the cover upward. Hold the hands together during the upward thrust to avoid catching an air pocket in the cover. When the cover is slightly higher than the objects to be covered, each operator snaps his hands outward. Each operator should follow through with his hands to guide the cover (Figure 3-84).

Step 5: Guide the cover over and down depending upon the position and length of the object to be covered (Figure 3-85).

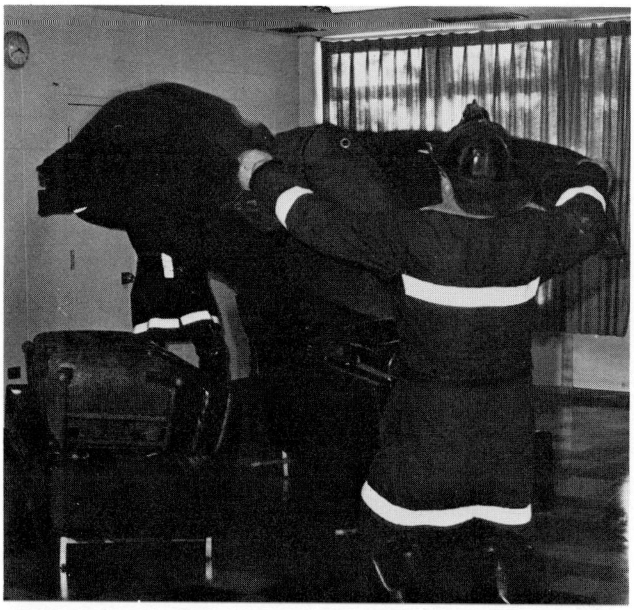

Figure 3-84 Continue in one movement snapping the cover upward, and then snapping the hands outward when higher than the objects to be covered.

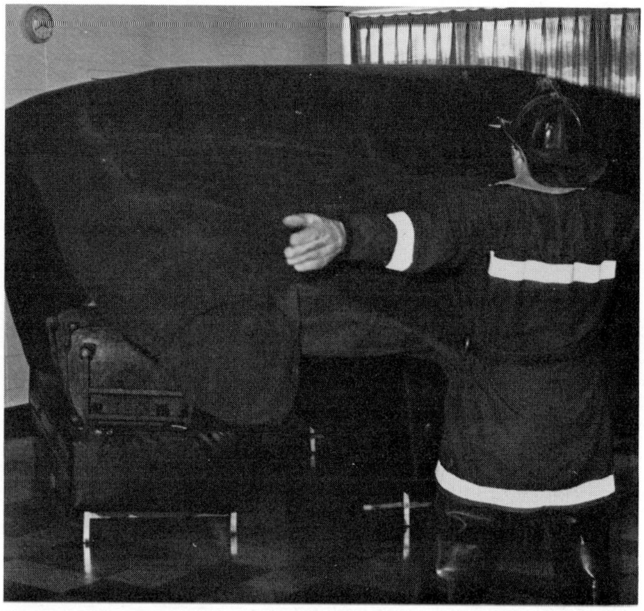

Figure 3-85 Guide the cover over and down as necessary to cover the objects.

SPREADING COVERS WITH THE CROSSOVER THROW

The crossover throw gets its name from the manner in which one edge of a salvage cover is crossed over the other edge prior to the throw. When the preparation is properly made, the underneath edge of the cover serves as a booster to push the upper edge over the object. Utilizing this arrangement, the pitch upward and over is made easier for both operators. The following steps are illustrated in Figures 3-86 to 3-91.

Step 1: Stretch the cover along one side of the object to be covered and bunch or roll both edges of the cover into the hands of each operator (Figure 3-86).

Step 2: The operator whose right arm is next to the object crosses his arms to hold the bunched edges (Figure 3-87).

Step 3: From the crossed arms position, bring the outside fold over the inside fold and cross at the middle. **Note:** This arrangement puts the cover in position for each operator to throw with his right hand. Each operator then moves the left hand toward the center of the cover one grommet. This shortens the length of the folded edge and aids in keeping it taut (Figure 3-88).

Figure 3-86 Stretch the cover and bunch or roll both edges.

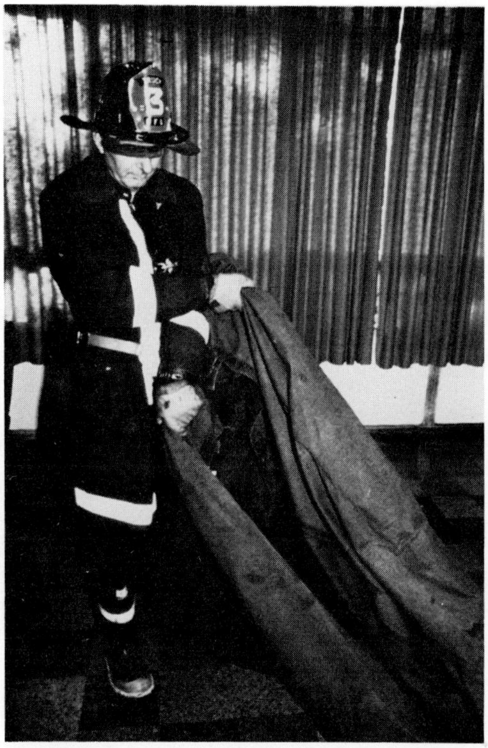

Figure 3-87 The operator whose right arm is next to the object crosses his arms to hold the bunched edges.

Figure 3-88 The outside fold is next brought over the inside fold.

Step 4: Pull tightly with the left hands and use this folded edge as a booster as each operator dips the cover toward the floor ready for the throw (Figure 3-89).

Step 5: Continue in one movement by pitching the cover upward with the left hands and toward the object with the right hands (Figure 3-90).

Step 6: When the folds from the right hand are over and across the object, bring the folds in the left hand down alongside (Figure 3-91).

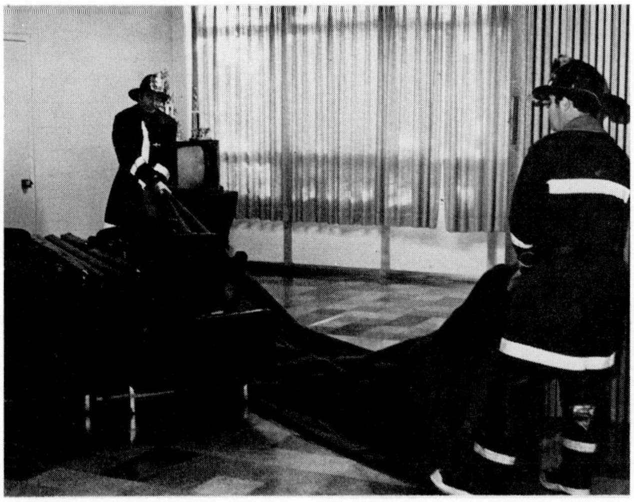

Figure 3-89 Pull tightly with the left hand and use this edge as a booster during the throw.

Figure 3-90 In a continuous movement snap the cover upward with the left hand and toward the object with the right hand.

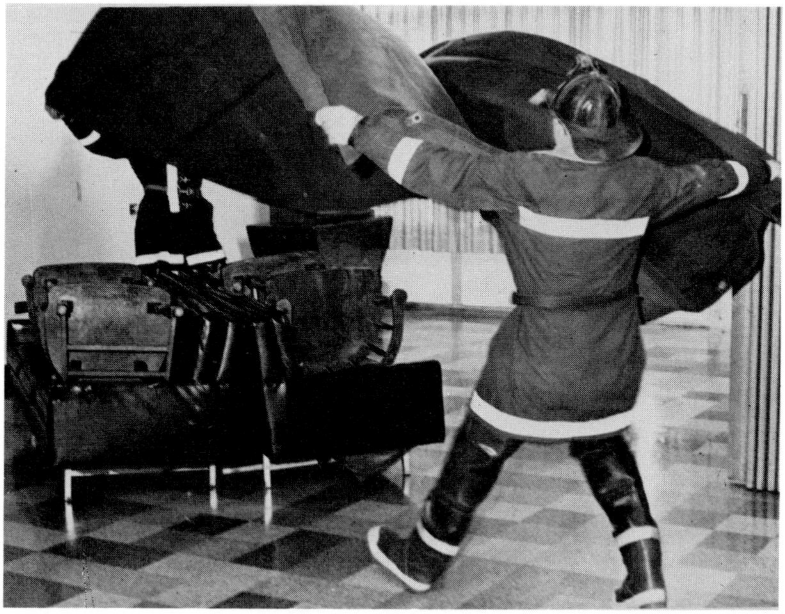

Figure 3-91 With the right hand folds over and across the object, bring the left hand folds down to complete the covering.

COVERING SHELVING

Unfortunately, not all shelving is constructed to permit proper covering during fires. Shelving is frequently built directly to the wall and it may be necessary to fasten the cover to the wall or top shelf in some manner. It is not advisable to use the shelving as a ladder to reach the top. This added weight applied to the front may cause the shelving to topple forward, causing injury to firefighters and also damaging the merchandise. When covering shelves, a short ladder should be placed at each end of the cover. After a salvage cover has been stretched between the two ladders, the firefighters carry one edge of the cover up their respective ladders to place it into position as illustrated in Figure 3-92.

Figure 3-92 Stretch the cover and carry it up short ladders when it is necessary to cover shelves.

The top edge of the cover can sometimes be draped over the top shelf and held in place by articles of sufficient weight which may be found nearby.

It may be difficult to keep water cascading down a wall from getting to the goods on a shelf from behind. The cover may be rolled on top of the shelf and forced between the wall and the shelving, or newsprint may be pressed against the dampened wall and angled into the cover allowing the water to follow it rather than the wall (Figure 3-93).

If the shelving to be covered is quite high, the cover may be opened the short way. To do this, the person with the cover on his shoulder takes either pair of single edges and gives the top single edge to his partner and grasps the other single edge himself. The cover is then opened the standard way and will be opened the short way as it unfolds. In some cases the best way to protect the merchandise on a shelf will be to remove it.

Figure 3-93 To keep water from getting behind shelving the cover may be rolled and forced between the wall and shelves, or newsprint used to form a path for the water to follow. The cover can be secured by weight with available items.

HANGING COVERS

A number of ways, as illustrated in Figure 3-94, can be used to hang covers to protect stock, form barriers, or reduce water damage. Hooks, nails through grommets, or cords can be used. When overlapping covers for extended protection, the overlap should be one grommet past the corner.

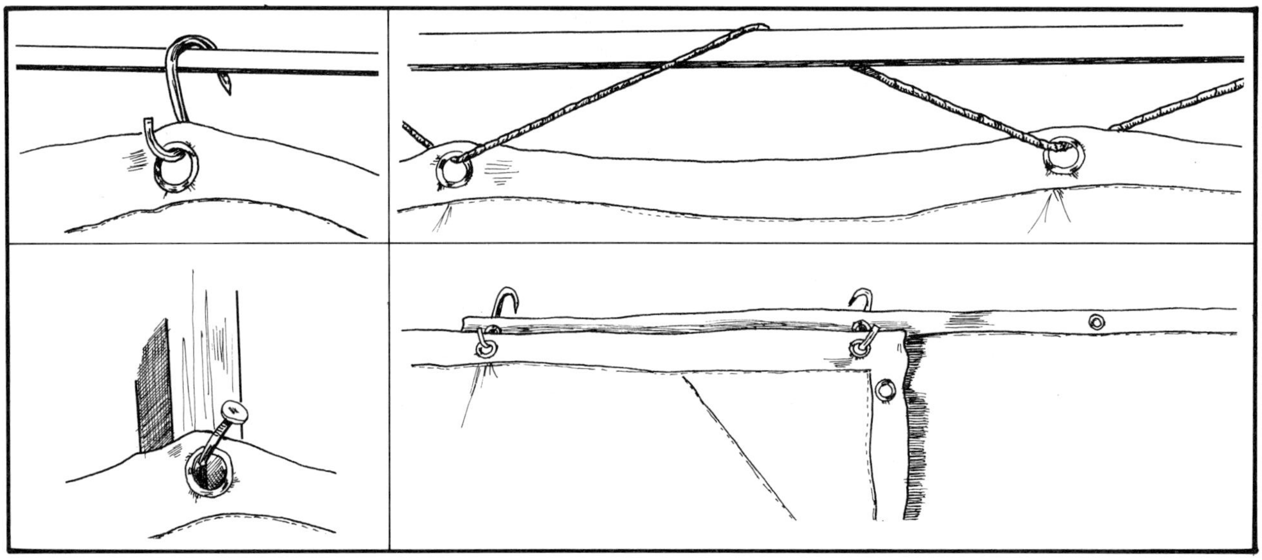

Figure 3-94 Numerous ways can be used to hang covers and the most suitable for a particular operation should be used.

REMOVING SALVAGE COVERS

When removing salvage covers, exercise caution and fold the cover onto itself to prevent snagging on sharp projections and to prevent damage to the materials that were covered. It is never advisable to pull a salvage cover from material. When a salvage cover is picked up after a fire, small amounts of debris and water usually remain on the cover. This debris and water should be contained within the cover while it is being carried outside so that parts of the building unaffected by the fire will not be soiled. The procedure is accomplished by two persons as follows (Figure 3-95).

While working from outside edges, systematically reduce the size of the cover by folding the corners to center. Reduce the size of the cover until one person can remove it without causing damage. After the cover has been removed it can be taken to the outside where the water and debris can be removed. By shaking the covers out and then folding them in front of the premises there can be no charges of removing valuable items in the salvage covers.

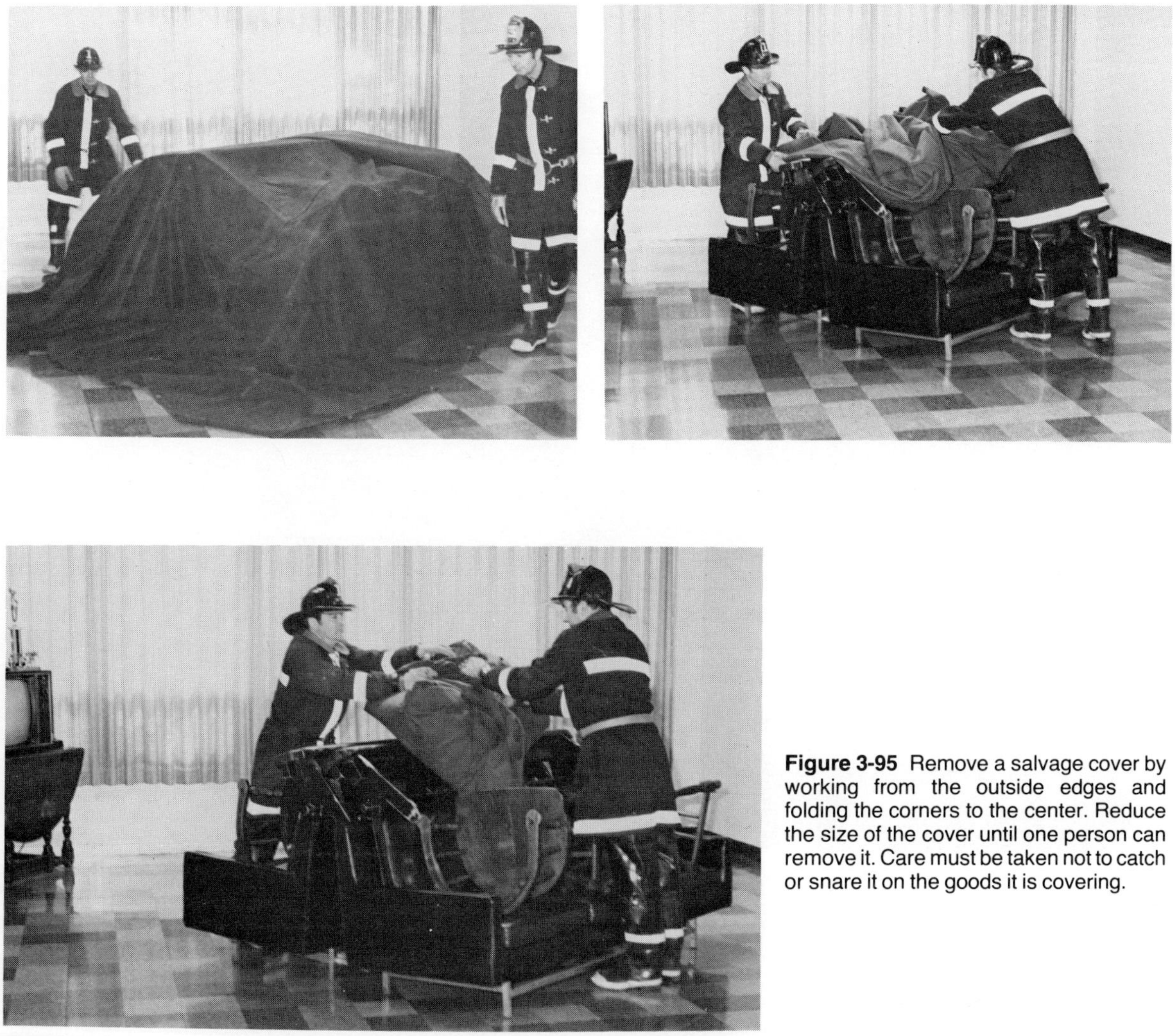

Figure 3-95 Remove a salvage cover by working from the outside edges and folding the corners to the center. Reduce the size of the cover until one person can remove it. Care must be taken not to catch or snare it on the goods it is covering.

Properly spread and hung covers will lessen smoke, water, dirt and heat damage keeping losses to a minimum.

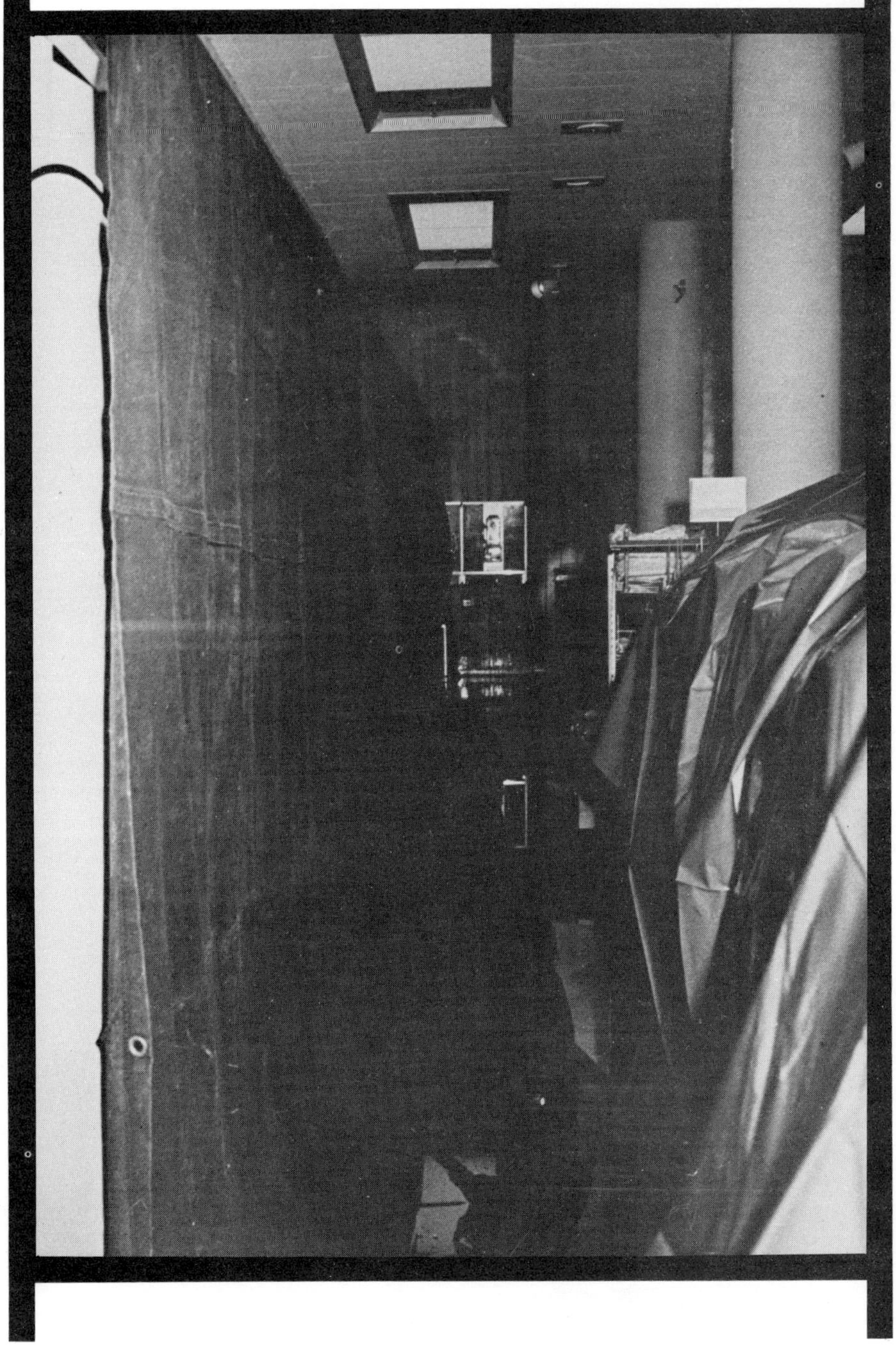

Improvising With Salvage Covers

Water will naturally tend to seek the lower levels of a building, but this natural condition can be controlled to reduce water damage. Controlling water runoff often involves the process of routing or directing water flow by utilizing stairways, elevator shafts, scuppers or other natural openings. Other methods of routing water from buildings may be accomplished by augering or cutting holes in floors, by constructing chutes and catchalls, or by mechanical devices.

Salvage covers have a number of applications beyond spreading and hanging to protect furnishings, fixtures and merchandise. They can be used to control and remove water when constructed into such devices as sumps, drains and basins. Covers can be used as dams to restrict the water flow. These uses reduce the area on the floor below that is subject to water leakage. The salvage job then becomes simpler since a smaller area will require covering.

In some cases covers will have to be spliced together to form larger chutes, or basins to cope with the greater quantities of water. Sprinkler discharge can also be limited by improvised cover usage. Smoke travel is limited by constructing curtains. Debris spread is restricted by protecting fireplaces while extinguishing a chimney fire. Properly covering hallways further reduces smoke and water damage.

IMPROVISED SUMP

The improvised sump is designed to serve a variety of purposes. It may be used to supply water to a pumper when threads are damaged or fittings are not available to make hydrant connections (Figure 4-1). The sump is an aid in supplying water from tankers in rural areas where discharge is limited and water has to be transported to the fire. The sump also can serve to trap and hold water in salvage operations.

Materials required in improvising a sump are a salvage cover, ladders, and rope or cord. Three ladders may be used, although a better procedure is to select a short 20 foot or 24 foot extension ladder and a matching roof ladder so that sections will nest between the beams of other sections. A pike pole may be used if a roof ladder is not available. The following steps describe how to make an improvised sump:

Figure 4-1 A sump can be improvised to allow the use of a hydrant that has non-standard or damaged threads or for which fittings are not available.

Step 1: Place the ladders and/or pike pole in a triangle, and lash the rungs and beams tightly together with rope or cord (Figure 4-2).

Step 2: Center a salvage cover on this framework, position the cover on the inside so that it fits snugly against the framework and the ground, and tuck the excess underneath. Use two covers and a leakproof seal if the framework is too large for a single cover (Figure 4-3).

Figure 4-2 Form a frame from ladders and/or pike poles properly secured.

Figure 4-3 Position a cover inside with the edges secured under the frame to build a sump for salvage work.

CATCH BASIN

Salvage operations sometimes require the hurried construction of devices to trap a considerable amount of water. The catch basin can be used for this purpose, although a suitable method for draining the basin must be considered. The fastest and usually the easiest way to improvise catch basins is described and illustrated below:

Step 1: Obtain suitable materials from the building contents such as chairs, tables, bedsprings, stepladders or stock to form a framework large enough to contain the anticipated amount of water (Figure 4-4).

Step 2: Spread a salvage cover to form the basin, tuck the loose edges underneath, and lash the components together to prevent spreading (Figure 4-5). Arrange for a suitable method of removing the water as the basin fills.

Figure 4-4 A catch basin can be framed using available items on the premises.

Figure 4-5 A cover is fitted to the frame, edges tucked under, and the basin secured to keep it from spreading while filling.

CATCHALL

A catchall is constructed from a salvage cover which has been placed on the floor to contain relatively small amounts of water. To control larger amounts of water a chute can be constructed to route the water from the catchall to the outside (Figure 4-6). Properly constructed catchalls will hold over a hundred gallons of water and can often save considerable time during salvage work. It may be necessary to construct a catchall under covered stock to contain the water coming from the protected goods (Figure 4-7). Since speed is important in controlling water, the cover should be placed in position as soon as possible. The side rolls can be started while opening the cover by turning the arms in several times as the cover is placed on the floor. Two operators are usually required to prepare a catchall in order to make more uniform rolls on all sides. The following steps required to make a catchall are illustrated in Figures 4-8 to 4-13.

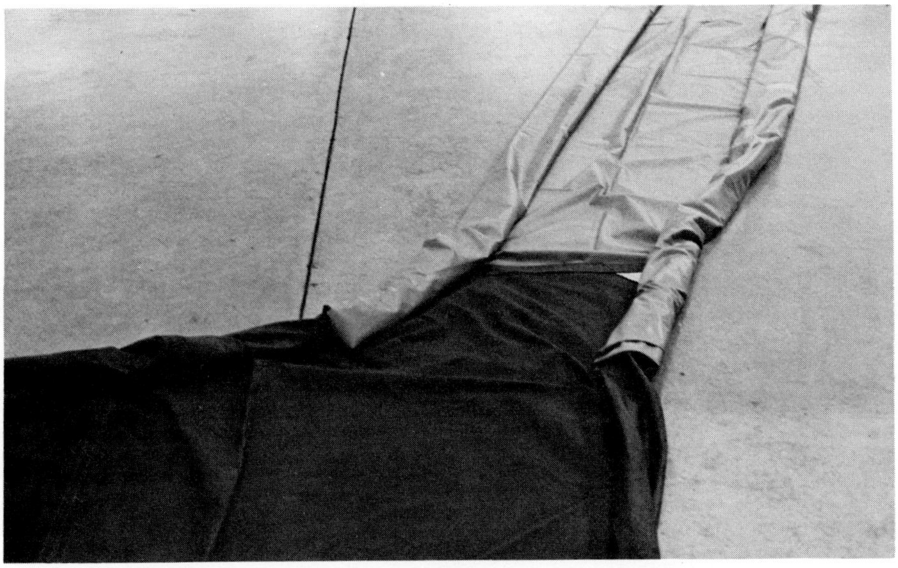

Figure 4-6 Catchalls may have to be continuously drained by a chute if they are collecting large quantities of water.

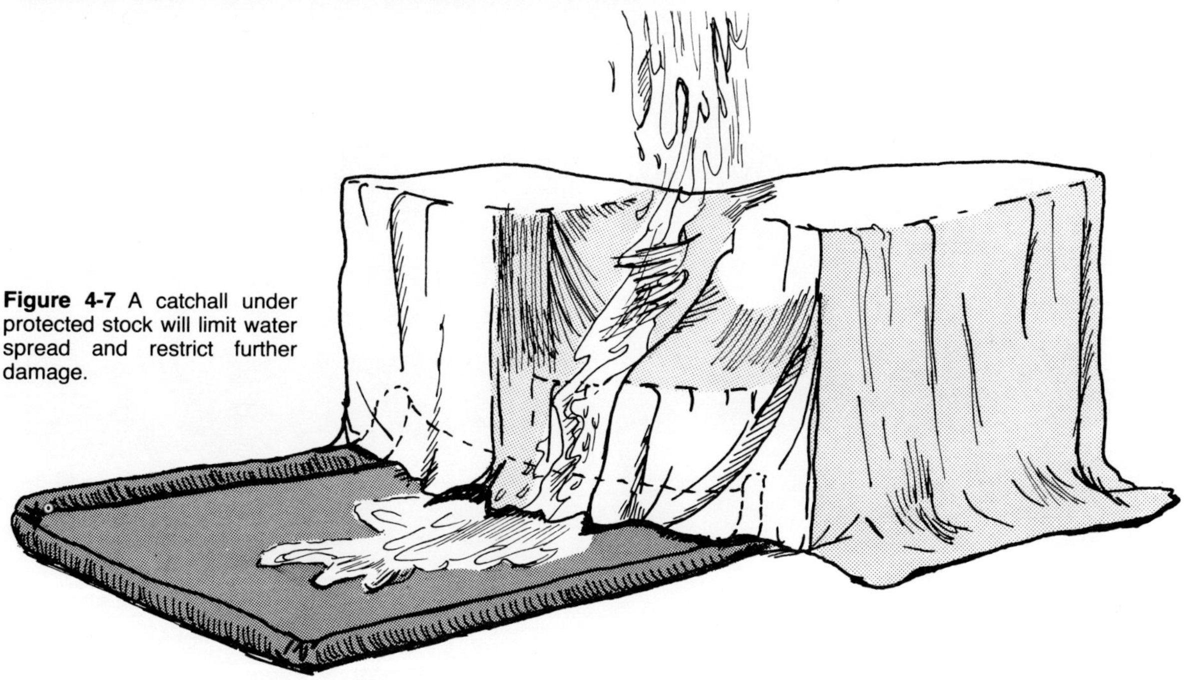

Figure 4-7 A catchall under protected stock will limit water spread and restrict further damage.

Step 1: With the cover spread upon the floor, roll the two sides inward approximately three feet (Figure 4-8).

Step 2: Lay the ends of the side rolls in at a 90-degree angle to form the corners of the basin (Figure 4-9).

Step 3: Roll one end into a tight roll on top of the side roll and form a projected ear. Using the outside hand to lift the edge roll, tuck the end roll under to lock the corners (Figure 4-10). An alternative to Steps 2 and 3 is to merely bend the end of the first rolled edge outward, and then roll the perpendicular side up to it and tuck the ear under the side to complete the corner. Moving the first edge in may be facilitated by using the toe (Figure 4-11).

Step 4: Roll the other end in a like manner and lock the corners. A completed catchall is shown in Figure 4-12.

Figure 4-8 A catchall is started by rolling each side inward about three feet.

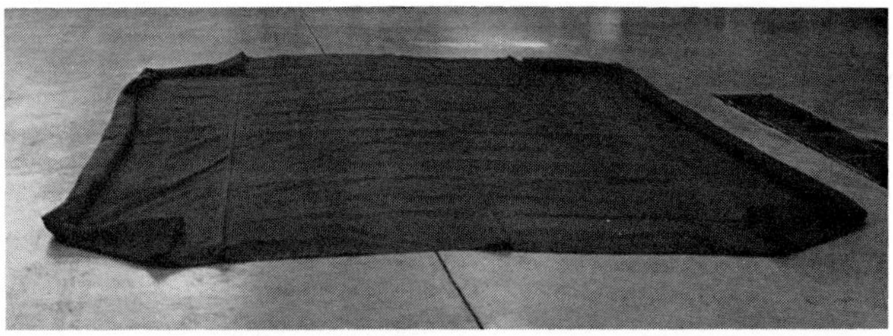

Figure 4-9 Place the end of the side rolls perpendicular to the sides to form the corners.

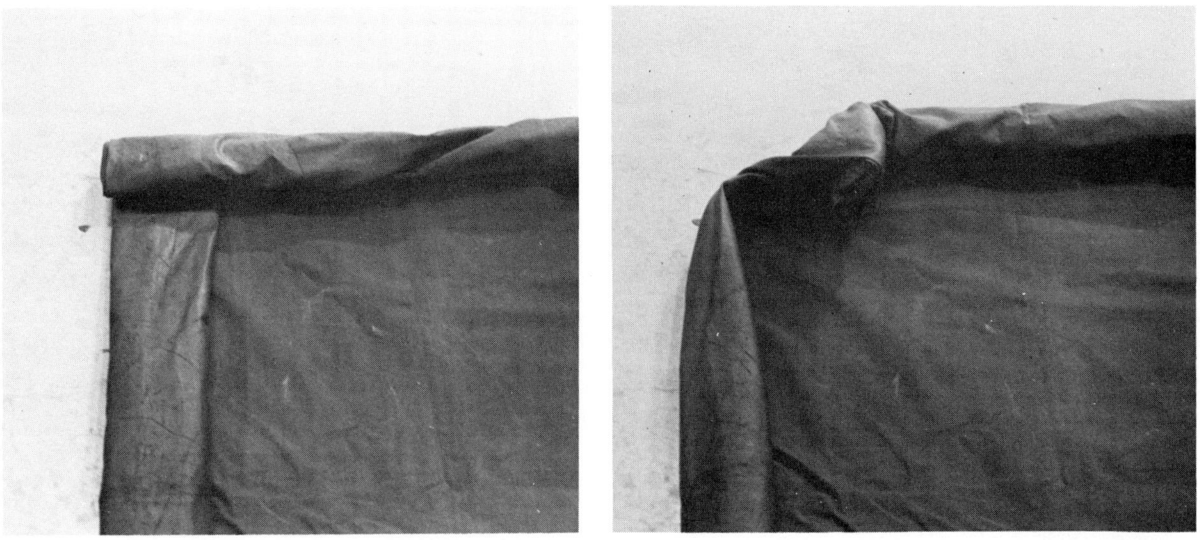

Figure 4-10 Roll one end into a tight roll on top of the side and then tuck the end of roll under the side wall to lock the corner.

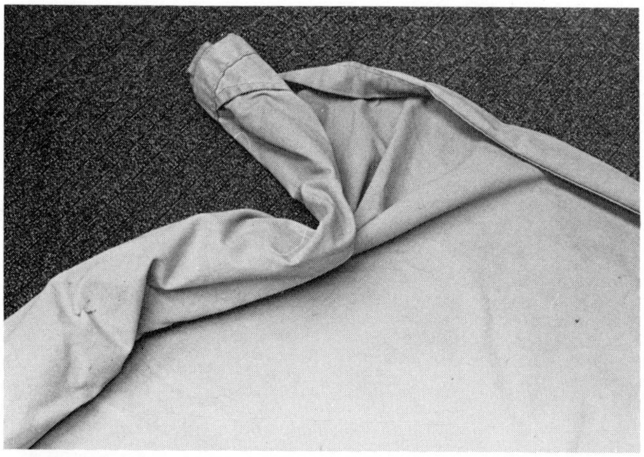

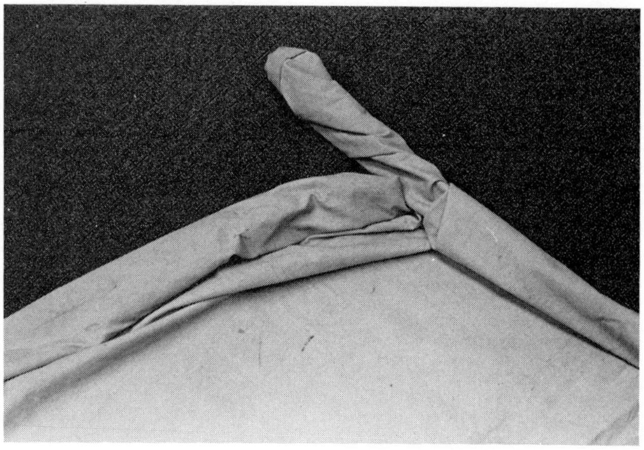

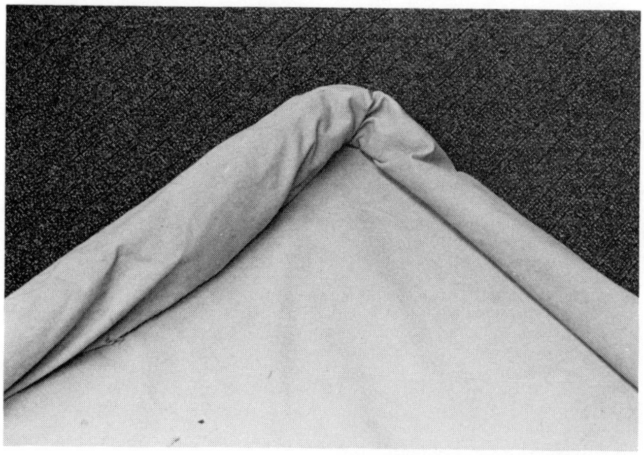

Figure 4-11 A corner can also be made by bending the side roll outward, rolling the side up to it, and tucking the ear under for completion.

Figure 4-12 A completed catchall ready for use.

Catch basins and catchalls can be easily emptied by use of a submersible pump. The pumps can be equipped to pump to one-eighth inch depths (Figure 4-13).

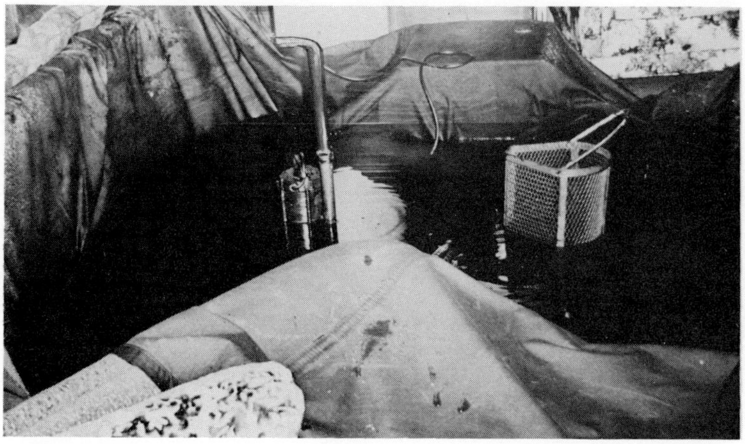

Figure 4-13 Submersible pumps work well for emptying catch basins and catchalls.

Alternate Catchalls

An alternate method of making a catchall is as follows:

Step 1: With the cover spread upon the floor, fold the ends in about two feet (Figure 4-14).

Step 2: Fold the sides in about one and one-half feet (Figure 4-15).

Step 3: Turn the tip of each corner in until it reaches the intersection of the folds (Figure 4-16).

Step 4: Fold or roll the four edges to complete the catchall keeping the walls as high as possible (Figures 4-17 and 4-18).

Figure 4-14 With the cover laid flat the ends are folded in two feet.

Figure 4-15 The sides are then folded in 1½ feet.

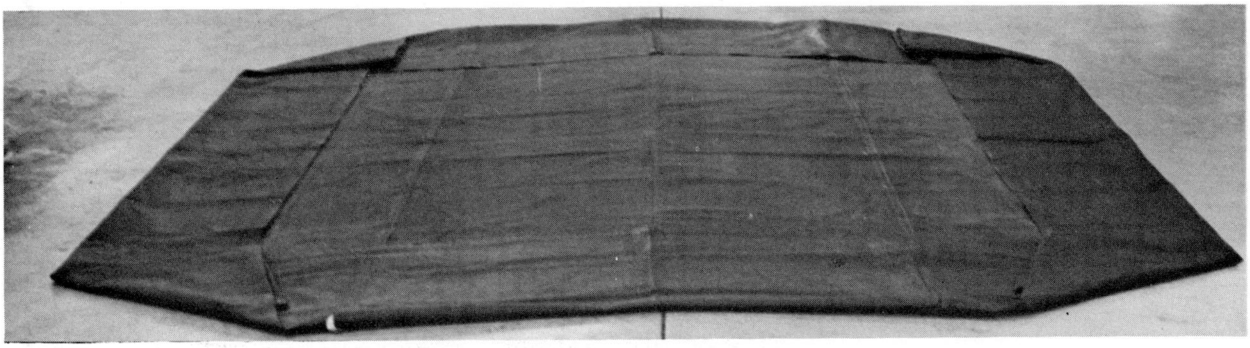

Figure 4-16 Turn the tip of each corner in until it reaches the intersection of the folds.

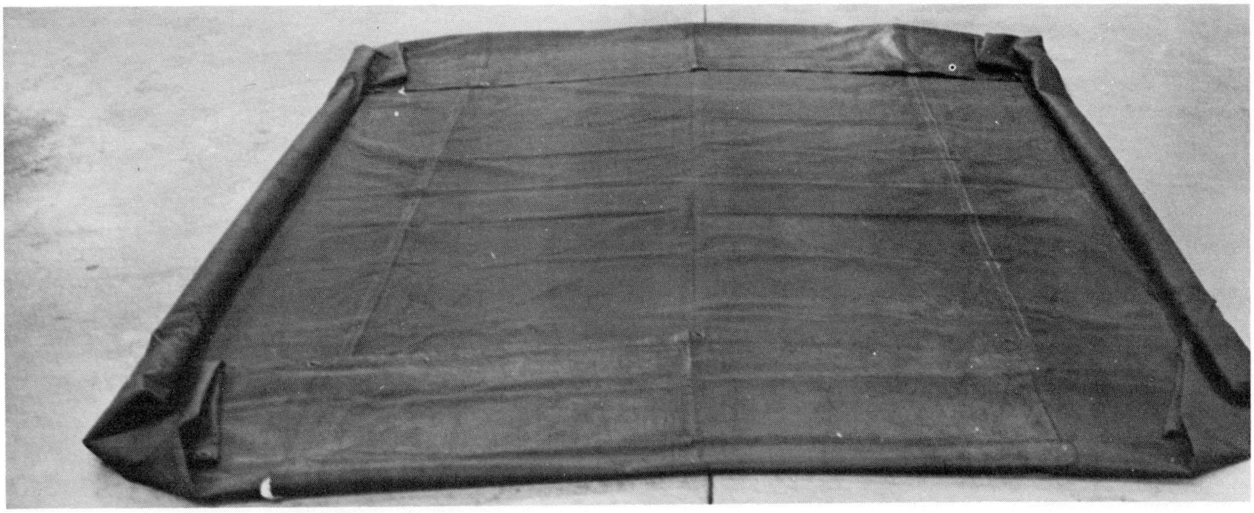

Figure 4-17 Fold or roll the four sides to complete the catchall.

Figure 4-18 Keep the walls as high as possible for maximum water control.

A quick method of making a catchall when four firefighters are available is as follows:

Step 1: Position a person at each corner and grasp the cover forming an ear (Figure 4-19).

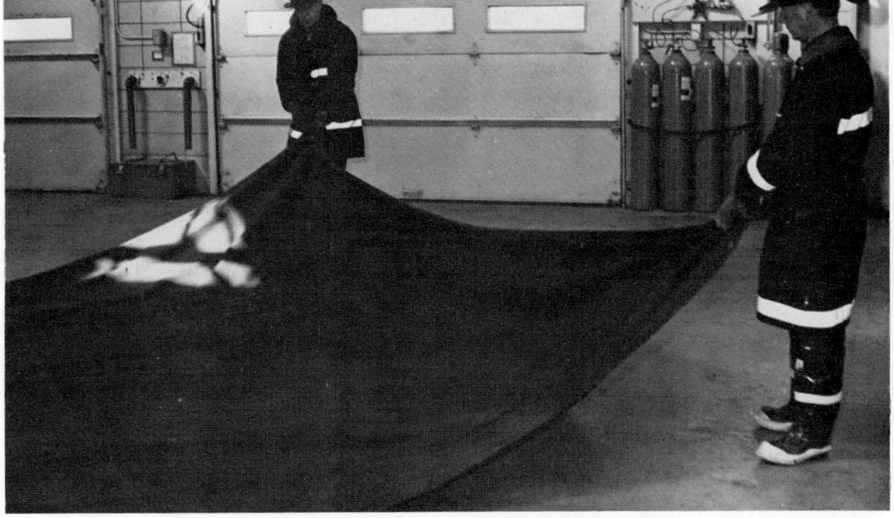

Figure 4-19 Position a person at each corner and form an ear from the cover corners.

Step 2: Each parallel pair of firefighters roll a side of the cover, then turn to the opposite corner and roll the perpendicular side (Figure 4-20).

Step 3: Tuck the ears in and roll to lock the corners. Place the catchall as needed (Figure 4-21).

Figure 4-20 Each parallel pair of firefighters roll a side of the cover, then turn 90°, and roll the perpendicular side.

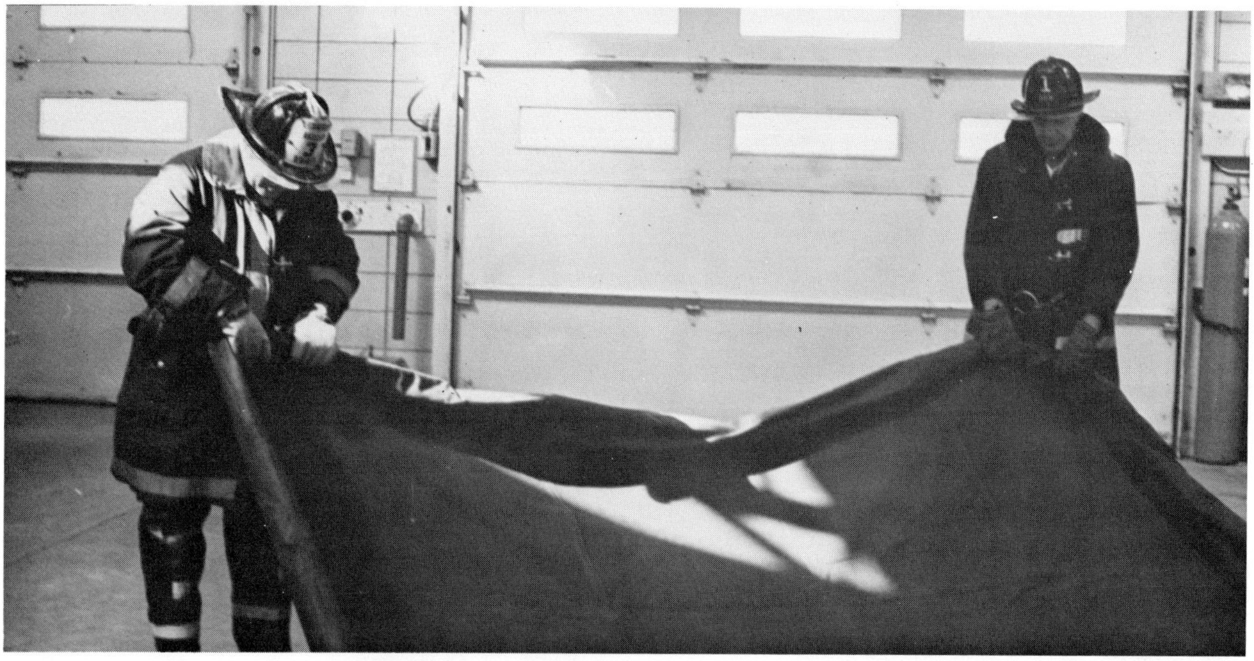

Figure 4-21 Tuck the ears in and roll as necessary to lock the corners.

CONSTRUCTING WATER CHUTES

One of the most practical means of removing water that comes through the ceiling from upper floors is by chutes. Water chutes may be constructed on the floor below to drain through windows or doors. Chutes must be placed high near the ceiling opening or the splash of a long drop will cause water damage. Some prepared chutes are available which are approximately ten feet long, but it may be more practical to make chutes by securing together two or more salvage covers. Effective water chutes can be made with two pike poles and a salvage cover. This operation is done by arranging the cover over two secured poles that reach outside a window or door as shown in Figure 4-22. Two alternate

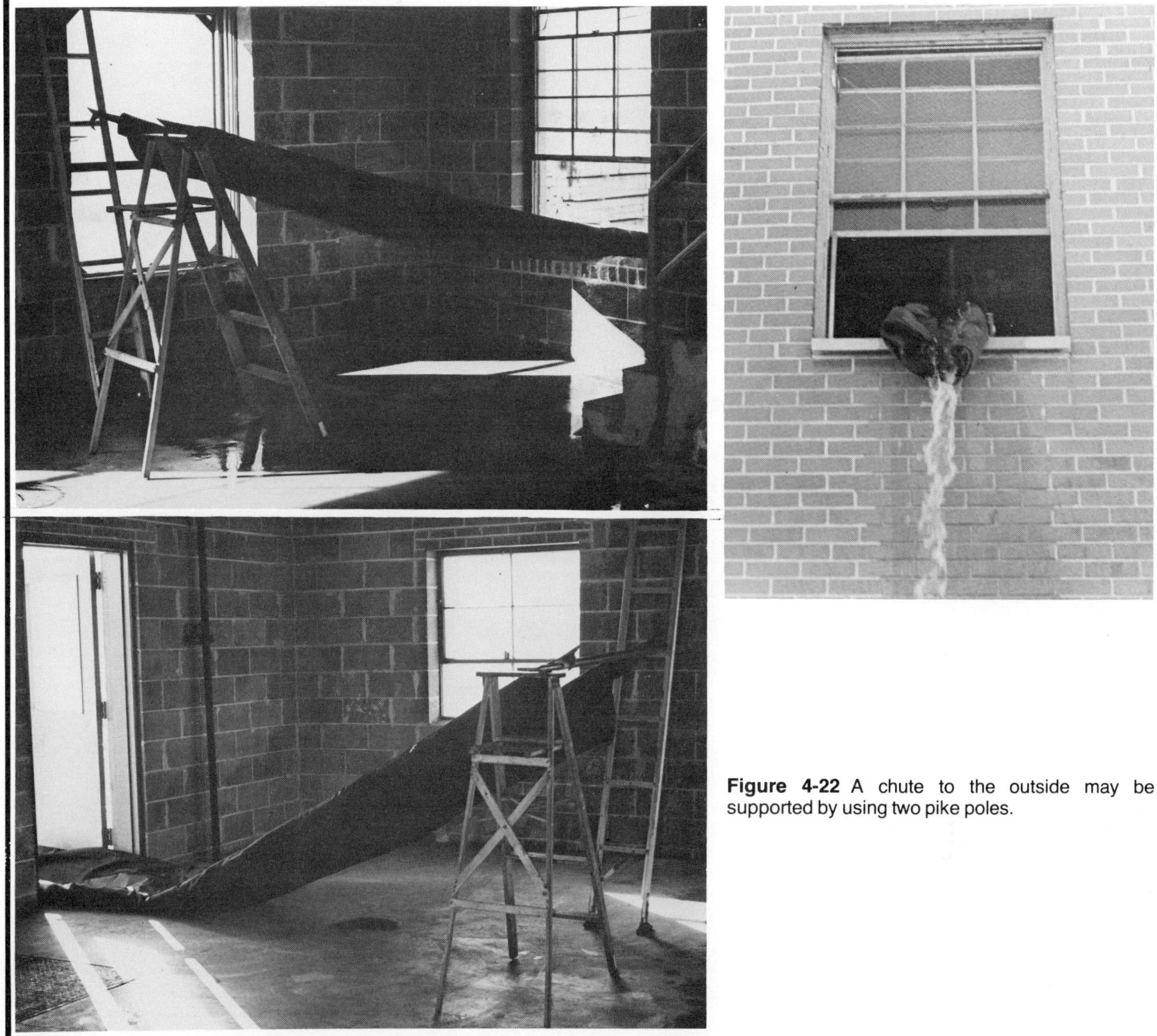

Figure 4-22 A chute to the outside may be supported by using two pike poles.

methods of forming a chute with pike poles are shown in Figure 4-23. A short chute can also be rapidly constructed by removing a door from its hinges, placing a cover on it, and properly securing it. Longer chutes can be made on ladders like the one shown in Figure 4-24. Chutes can be supported in a number of ways using available equipment such as pike poles or ladders (A-frame, step, or locked together). A salvage cover is partially unfolded in order to roll the edges which are placed inside the ladder beams. Note that when the rolled edges are placed underneath the cover to prevent loosening, the weight of the water in the trough tends to tighten the rolls. On high beam ladders the rolls do not have to be inverted since the beams will hold them.

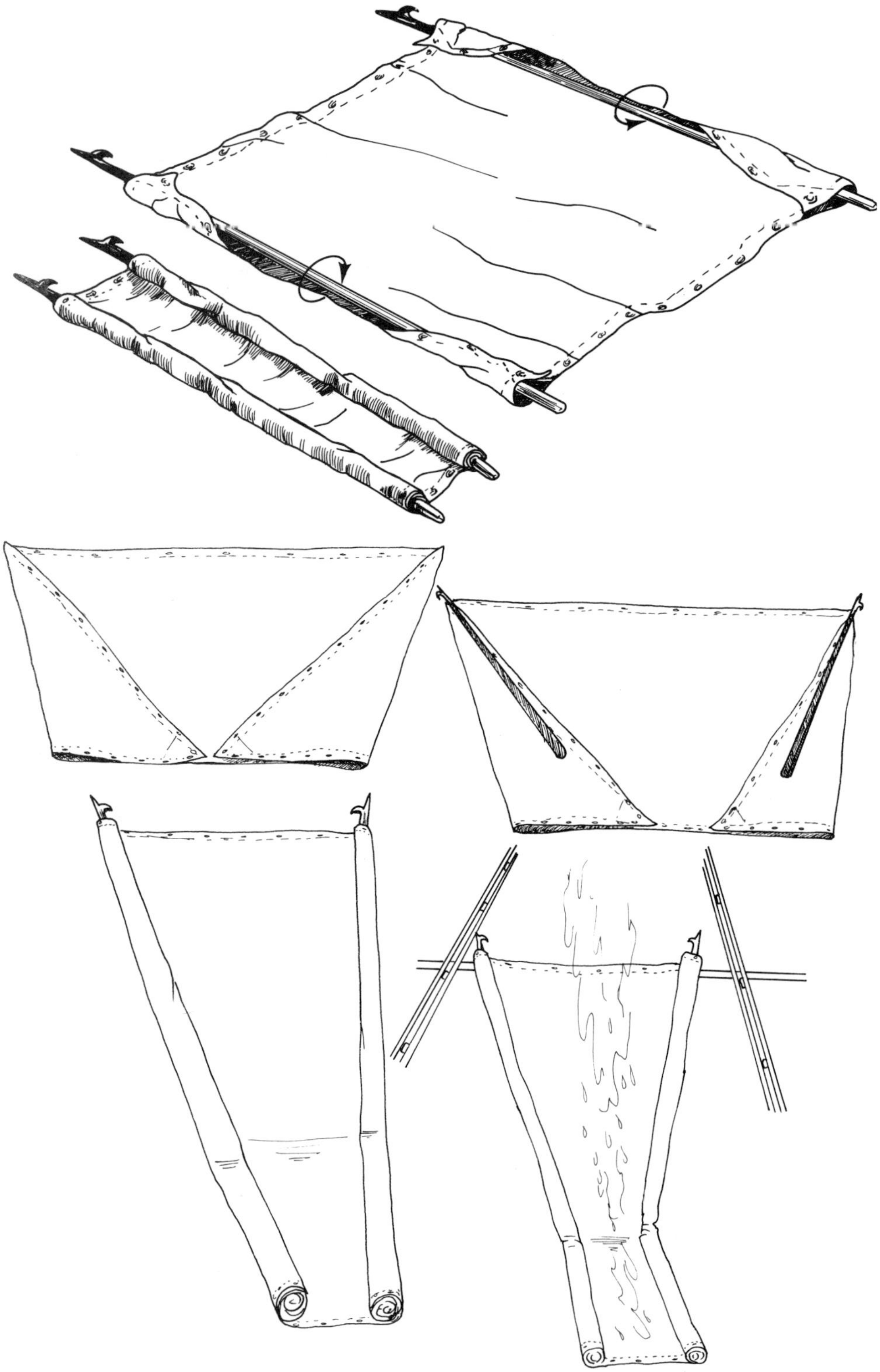

Figure 4-23 Two alternative methods of constructing water chutes using pike poles.

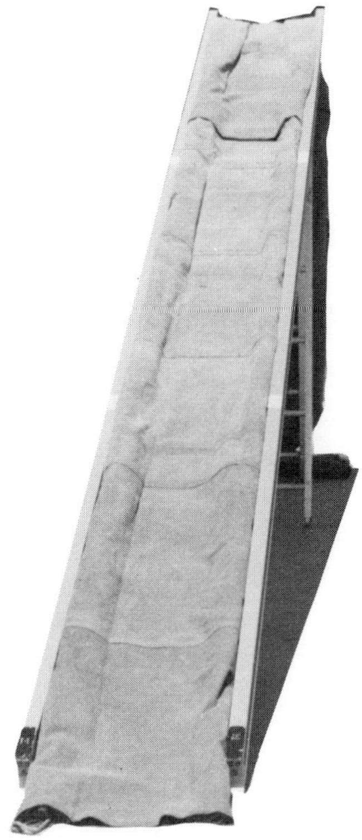

Figure 4-24 Covers can be spliced to form a longer chute supported by a ladder.

The proper drainage angle can be maintained in a number of ways using ladders, pike poles, "S" hooks and cord, or even the material found on the premises. Salvage work requires ingenuity. Crossed ladders as in Figure 4-25 would feature a pike pole through common rungs to hold up the pike poles supporting the rolled chute, or two pike poles may be crossed and the salvage cover chute attached to the upper hooks to provide good angles for drainage. A rolled cover supported by a ladder can be unrolled at the top and widely spread for better water collection. "S" hooks through the grommets may be used with cord and other "S" hooks to hold the top up and spread.

Two methods by which water chutes may be applied on stairways are shown. A stair banister being used to support one side of the cover is illustrated in Figure 4-26. Both sides of the cover rolled to form a dike is illustrated in Figure 4-27. Notice that the rolled edges are again placed on the underneath side to help prevent the cover from unrolling. This method of chuting can also be used on escalators.

Figure 4-25 A ladder is used here to support a chute. Various other methods are available and limited only by a firefighter's ingenuity.

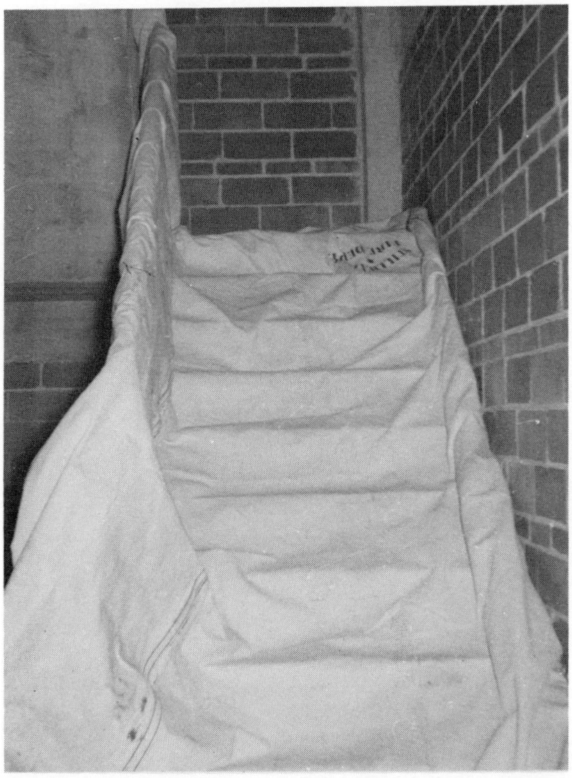

Figure 4-26 A stairway banister rail can be used to support a chute on a stairway.

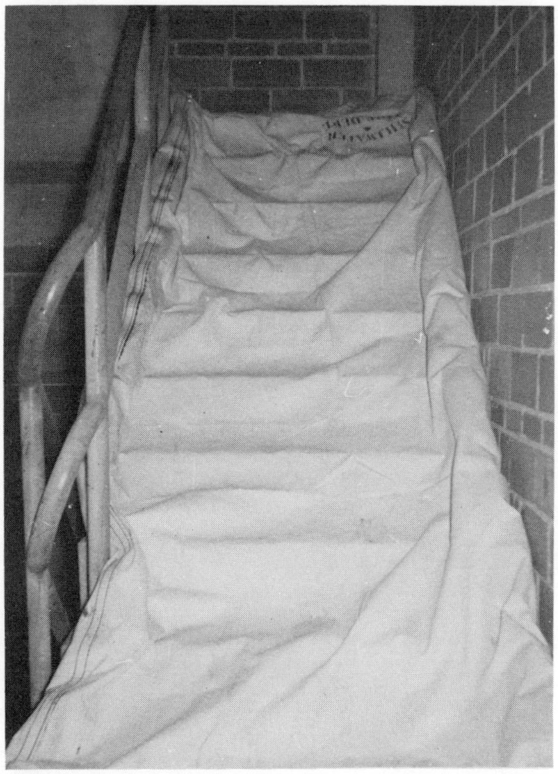

Figure 4-27 A stairway can also be covered by rolling the cover edges, flipping the cover over, and forming a chute working from the bottom up the stairs.

SPLICING COVERS

When objects that are too large for one cover need to be covered, or when long chutes or catchalls need to be made, it is necessary to splice covers with watertight joints. Also, when water must be routed from a catchall into a chute, the water must be prevented from running back under the catchall. The steps for splicing covers are explained and illustrated below, first to drain a catchall:

Step 1: Unroll one corner of catchall and spread (Figure 4-28).

Step 2: Unroll one end of chute and fold it back over itself about 12 inches (Figure 4-29).

Step 3: Lift corner of catchall and unfold chute beneath.

Step 4: Roll sides of catchall corner and end of chute back together (Figure 4-30).

Figure 4-28 Unroll one corner of the catchall.

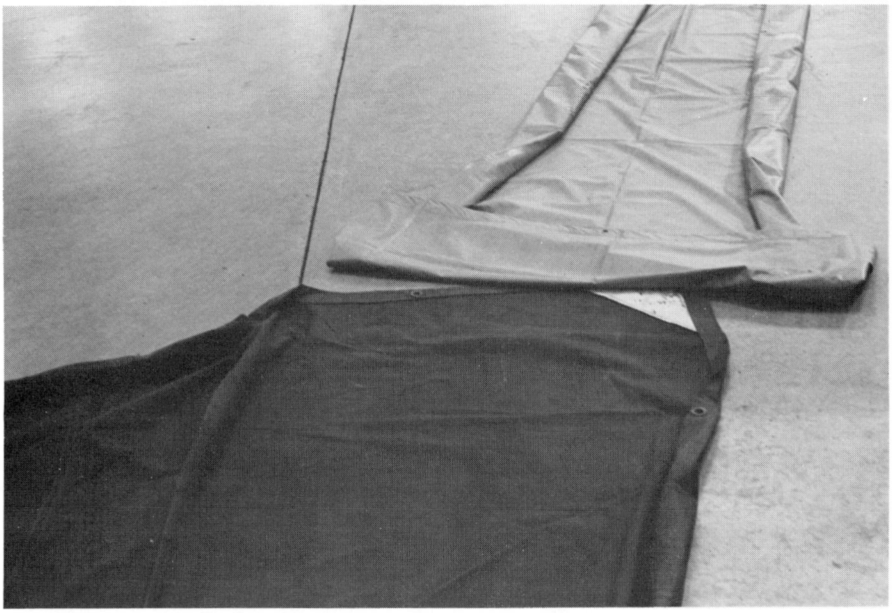

Figure 4-29 Fold back the unrolled end of the chute 12 inches.

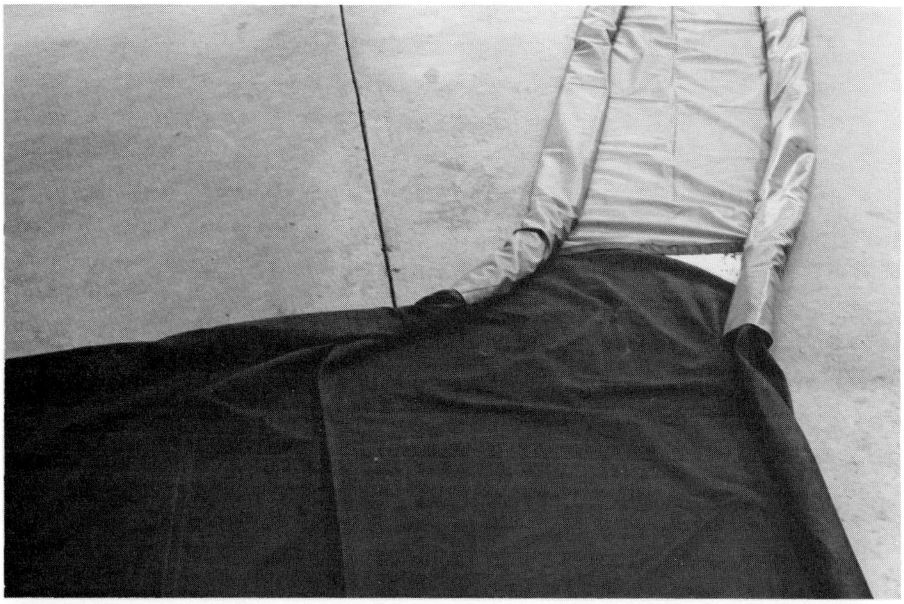

Figure 4-30 Lift the corner of the catchall and unfold the chute below it; then roll the sides of catchall and chute together.

When a long chute is necessary and water must flow over a joint, it may be folded as described in the following steps:

Step 1: Unroll the edges of the uppermost cover to expose the end and fold this end back upon itself about 20 inches (Figure 4-31).

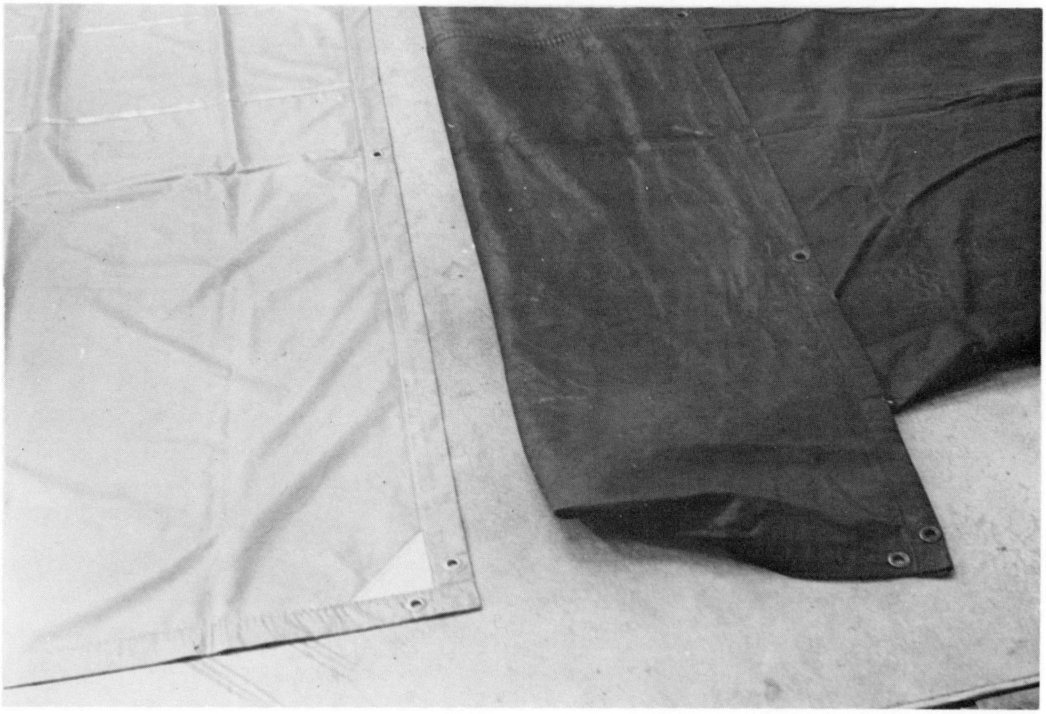

Figure 4-31 Upper cover unrolled and folded back on itself about 20 inches.

Step 2: Lap the second or underneath cover over this fold about ten inches (Figure 4-32) and fold the remaining ten inches of the uppermost cover back over the second cover (Figure 4-33).

Step 3: Fold the entire joint of both cover ends down over the underneath (second) cover (Figure 4-34).

Step 4: Press the joint smooth so that water will flow freely over the lap joint (Figure 4-35).

Step 5: Roll the edges of the second cover in line with the rolled edges of the uppermost cover for the completed splice (Figure 4-36).

Figure 4-32 Lower cover lapped ten inches over upper cover.

Figure 4-33 Remaining ten inches of upper cover is folded over lower cover.

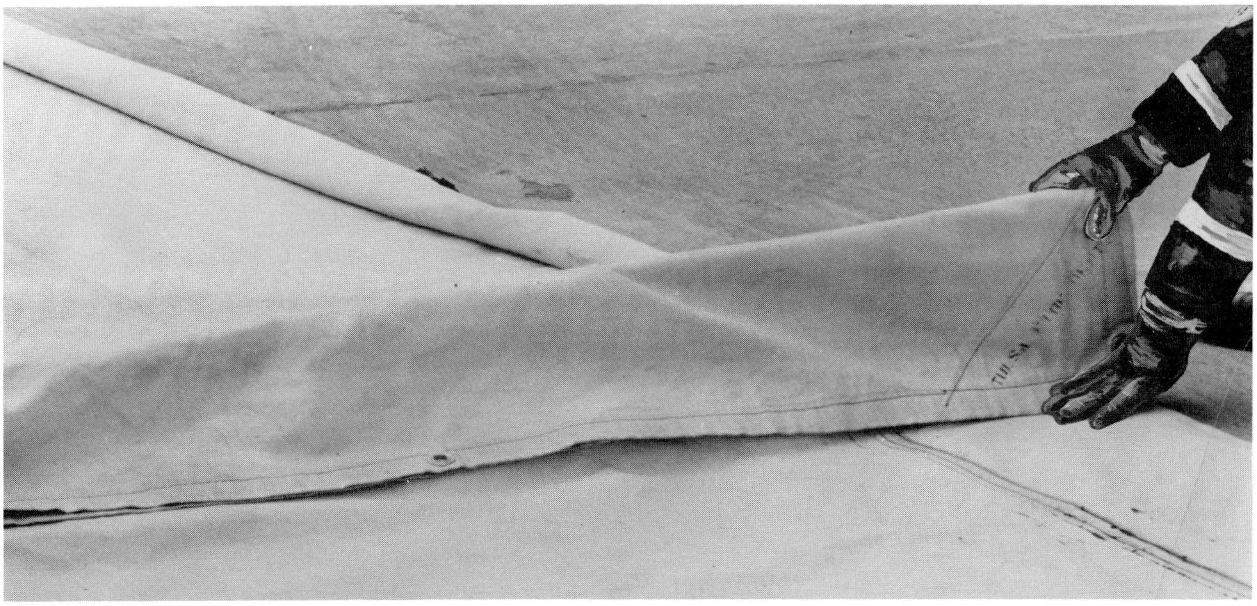

Figure 4-34 Fold the joint of the two covers down on the lower cover.

Figure 4-35 Press the joint smooth to assure free water flow.

Figure 4-36 Roll the edges of the two covers including the splice back forming a continuous sidewall.

If the joint is to hold or route water, fold the edge of one cover back over itself about 18 inches toward the direction the water is coming from and bring the edge of the second cover over and even with it. Then roll or fold the two covers tightly together. The proper method of folding is to make two eight-inch folds with both covers back toward the second cover or in the direction of water flow. The folds should be pressed flat.

A joint of this type may be made outside the building where space is available and then the sides rolled in and the ends folded to the center for carrying. Covers made up this way can be used for stairways, hallways, aisles or large open areas like a bowling alley.

CONTROLLING SMOKE, DEBRIS AND WATER

Salvage covers can be used effectively as smoke curtains, debris curtains during chimney fires, as debris carriers, to control water by covering hallways and sprinklers, or as dikes.

Smoke Curtains

Smoke can be controlled in most open stairways by using one or two salvage covers as curtains at each landing. This method makes smoke removal possible through windows or other vents, or through openings made above the stairways, without allowing smoke to enter other floor levels. Curtains may also be used to prevent smoke from entering exit stairways or to block smoke from passing into uninvolved sections of the building (Figure 4-37). All of the basic principles of ventilation should apply when smoke curtains are used. Smoke curtains can be secured in place in dwellings with an extendable athletic bar or smoke ejector holder which is designed to fit across a door opening.

Once on the fireground, the efficient salvage officer will use whatever is available to assist in salvage. Improvising on the scene will lead to many damage saving operations by controlling dirt, heat, smoke and water.

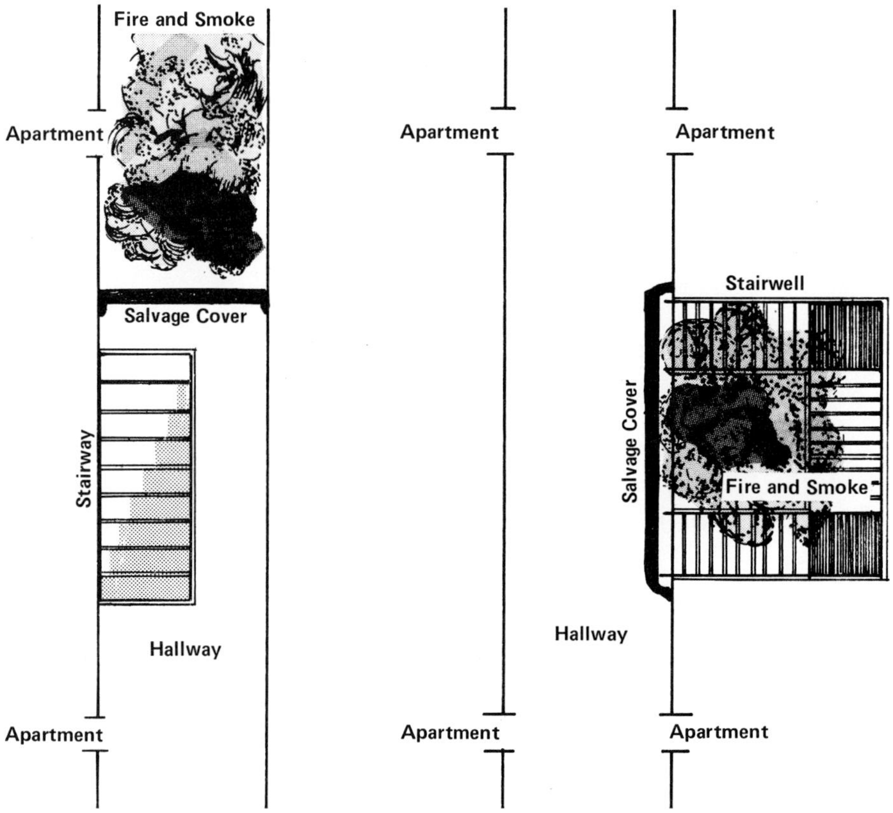

Figure 4-37 Salvage covers can be used to limit the vertical and horizontal spread of heat and smoke by curtaining off areas such as hallway or stairwell.

Protection During Chimney Fires

At times the interior of a room must be protected from the ashes and debris found in a fireplace. Usually this will occur when men are controlling a chimney fire from the roof or chimney top.

One method of chimney fire control requiring the use of a canvas to cover the fireplace opening involves reducing the flames in the fireplace with a minimum of water spray and then applying a water spray down from the top of the chimney. After the flames in the fireplace have been extinguished, but before water is applied from the top, a canvas should be hung or supported over the fireplace opening to prevent smoke and ashes from drifting into the room. Many fire departments carry a "chimney-fire canvas," six feet to eight feet square, for this purpose.

When a mantelpiece is present, the canvas can be hung in front of the fireplace by placing weights such as books on the part of the canvas laid on the mantel. In the absence of a mantel, the canvas will have to be held or propped against the face of the fireplace.

Hallway Covers

Covers carried in a two-person fold can be quickly used to cover a hallway. The cover is stretched and the corners gripped in each hand. The cover is pulled taut and the hands turned in a circular motion ↷↶ to take up the excess width and form side rolls. The rolls are then placed against the walls of the hallway covering a 14- to 18-foot length. This operation although not neat looking takes only seconds to cover and protect the hall floor (Figure 4-38).

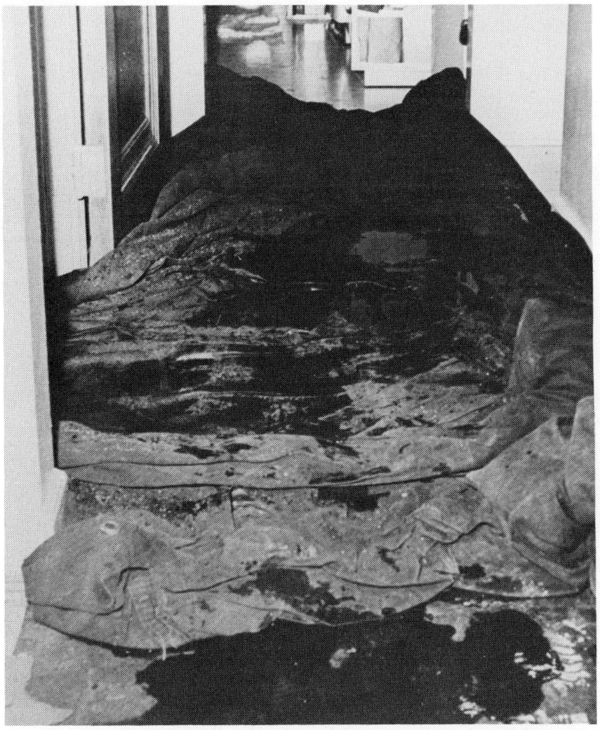

Figure 4-38 Hallway quickly covered using a hallway throw.

Dikes

Dikes or dams can be built across door openings to limit the water spread. A cover should be opened half way, placed in the opening with the open edges away from the water, and folded covers stacked inside to form a dike. The original cover is then folded over the piled covers completing the dike (Figure 4-39). Folded covers can be used to block doorways or form dikes to control water in a number of operations.

Figure 4-39 Folded cover can be placed in an opened cover to form a dike in a doorway, the cover is then folded over the stacked covers and the dike completed.

Sprinkler Operations

When a high sprinkler head cannot be reached with a stopper, a catchall or catch basin should be quickly built below it. While the catchall or catch basin is being constructed under the open sprinkler head, a cover can be stretched and held above the floor to hold the discharge. The cover is then placed in the basin or catchall (Figure 4-40).

Figure 4-40 A cover can be stretched to collect the discharge from an operating sprinkler head while a catchall is being constructed below.

If a sprinkler stopper is not available, a folded cover can be draped over the head and a catchall built below. The water will drop straight down and into the catchall limiting the water spread.

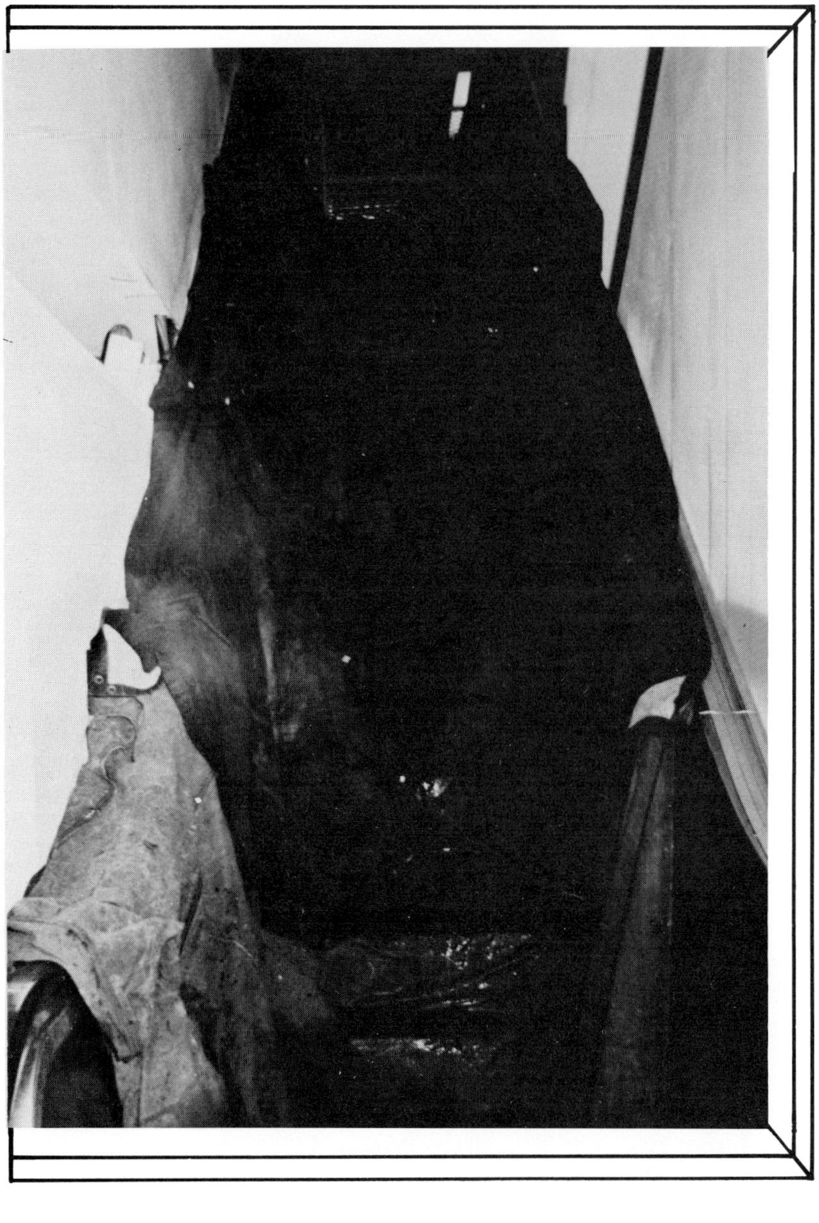

Improvising with salvage equipment can protect high value merchandise and equipment. Adequately protecting this escalator prevented serious damage and also provided a chute to remove water limiting water damage on the floor above.

5

Handling Water In Buildings

After the protection of building contents, the concern over excess water is normally the next most important salvage function. Not only is water a concern because of its weight and its natural tendency to work downward in a building, but it is a concern as it accumulates or is absorbed in areas that are without adequate drainage. This accumulation of water can damage household items, stored goods, machinery and utilities as well as the structure.

A basic premise of fire fighting is that all personnel should be aware of the additional weight placed in a building by water. An idealistic hope is that all water applied on a fire will be converted to steam during extinguishment, but unfortunately this is not the case. Where none of the water is being vaporized because it is not being applied on or in an area of sufficient heat, and the application is from a 250 gpm nozzle, simple mathematics indicate that just over a ton of water per minute is being added to the building. Obviously, continued application of a number of streams into a building, with most of the water remaining in liquid form, will rapidly increase the weight.

In most buildings accumulated water will work its way to the lowest level. Water in lower levels could endanger machinery, utilities, stored merchandise, heating, ventilating and elevator equipment.

ROUTING WATER

There are numerous avenues that can be used to route water. Use of floor drains and sewer openings cause the least additional damage. Stairways and chutes to the outside are convenient. Scuppers, where available, will clear floors, or elevator and other shafts can be used. Walls may be breached, holes cut in floors, or pumps used in addition to improvising methods.

Methods of removing water from basements or shafts involve either draining or pumping. Drains in basements are often clogged by debris or covered with stored merchandise. If drains are controlled by valves or if the basement has a switch for manually activating a pump, these should be noted during preplanning and indicated by signs suspended from the ceiling. The drains should also be indicated so that they may be cleaned as they become clogged. Such drains and signs should be discussed during pre-fire inspections.

Excess water on upper floors must be collected and removed in a way that will do as little damage as possible. Here the ability, knowledge and ingenuity of firefighters may be taxed to the utmost because of the many different conditions found in this area of salvage work. A complete knowledge of all tools and salvage equipment and the techniques of using them is essential.

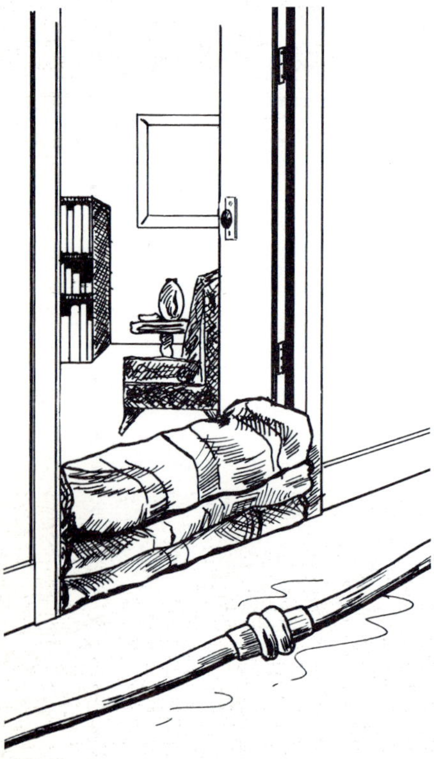

Figure 5-1 Dikes can be made of salvage covers to channel water and keep it from spreading into adjacent rooms.

Since water seeping through a floor will have a tendency to follow a lighting fixture to the floor below, it is advisable to remove the fixture and place a hole next to the electrical box and in the floor over it. After water removal, the fixture can be replaced covering the hole in the ceiling.

Water following a pipe through a floor can be handled by tightly binding a salvage cover around the pipe with twine and bagging the floor.

When feasible, catchalls should be constructed to contain the water. If, however, the amount of water is large or the flow is continuous, the water will have to be channeled out of the building with chutes or dikes. Chutes may be constructed of salvage covers and a framework either of ladders, a stairway, or contents of the building. Dikes may be made of salvage covers (Figure 5-1) or sawdust. Sawdust is also used to absorb standing water that cannot be channeled to a drain.

Another method of draining water is through the soil pipes installed in a building. This can be done either by removing a toilet bowl, a cleanout plug, or by breaking a vertical soil pipe at floor level. Remember that sewer gas leaking from a broken soil pipe can be a problem.Toilet bowls can easily be lifted after using a wrench to remove the fastening devices holding the base to the floor (Figure 5-2). By rocking the bowl after the fasteners have been removed, the wax or putty seal can be broken and the bowl

Figure 5-2 Toilet bowls can be easily lifted to expose the floor level soil pipe for removing water.

lifted out of place, revealing the flush soil pipe opening. Early removal of the water is of essence. Therefore, if the water is deep, the fasteners rusted, or there is any difficulty in removing the toilet, it is better to break the toilet causing less damage than to allow the water to continue to build up and spread.

A wrench will be needed to remove cleanout plugs in soil pipes (Figure 5-3). Tapping the edge of the bell may help if the plugs are hard to start. If these methods fail, a soil pipe extending from the floor can be broken off at floor level to drain the water (Figure 5-4). When any of these procedures are followed, screens or drain guards should be used to prevent large pieces of debris from entering the drain pipes.

Figure 5-3 An exposed soil pipe will require a wrench to remove the cleanout plug and possibly some tapping to loosen the plug. Also a screen should be used to prevent large pieces of debris from entering the drain pipe.

Figure 5-4 A soil pipe can also be broken at the floor level with a sledge to form a drain.

One of the objectives of salvage work is to remove excess water to the exterior of a structure. Sometimes this can be accomplished by the use of sewer pipes, catch basins or drip tubs, stairway chutes, or other means. But if none of these methods is practical during a particular instance, the most effective method then involves cutting holes in upper floors and diverting water through window drains or chutes. Before a hole is cut in a floor, merchandise or furnishings should be covered and chutes constructed in the area below. The hole should be placed so that the water will flow onto a chute that extends through an exterior window or door. The hole in the ceiling above the chute should be cut first, and it should be somewhat larger than the hole in the floor above so that water will not accumulate between the floor and the ceiling, as illustrated in Figure 5-5.

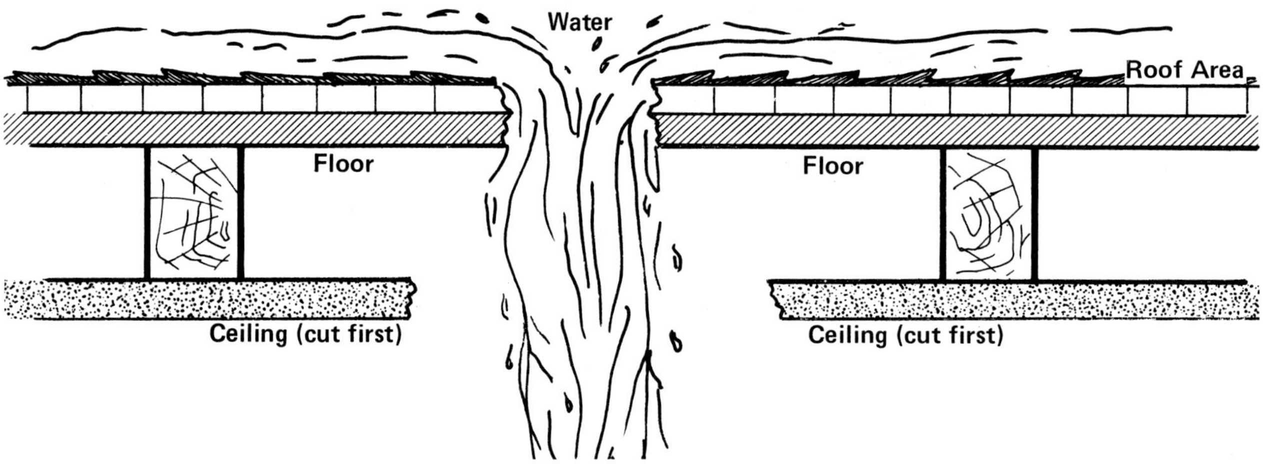

Figure 5-5 When cutting holes in floors, open the ceiling below first with a somewhat larger hole to alleviate any entrapment of water between the ceiling and floor.

REMOVING WATER WITH PUMPS

It is not uncommon for various quantities of water to partially fill the area below ground level. These sumps or basements may be provided with a drain into a domestic sewer, but the drain may be clogged with debris. Fire departments are not normally prepared to remove water from a large number of properties when storms or floods cause excessive damage. Regular fire department pumpers should *never* be used for pumping water from buildings for these three reasons: (1) fire apparatus is too expensive to be used for this purpose and relatively inexpensive portable pumps are available; (2) the debris-laden water is likely to damage fire department pumps (diaphragm and submersible pumps are designed for this, Figure 5-6); and (3) fire apparatus should be kept available for fire fighting. It is the responsibility of the fire department, however, to remove from buildings water which has been used to extinguish the fire. Portable water pumps and diaphragm pumps of various capacities, which are powered by small internal combustion engines, are ideal for this purpose. Two examples of portable pumps are shown in Figure 5-7. The electric submersible pump can also be used for dewatering. These are lightweight and easy to use. Capacities can be obtained to over 350 gpm (Figure 5-8).

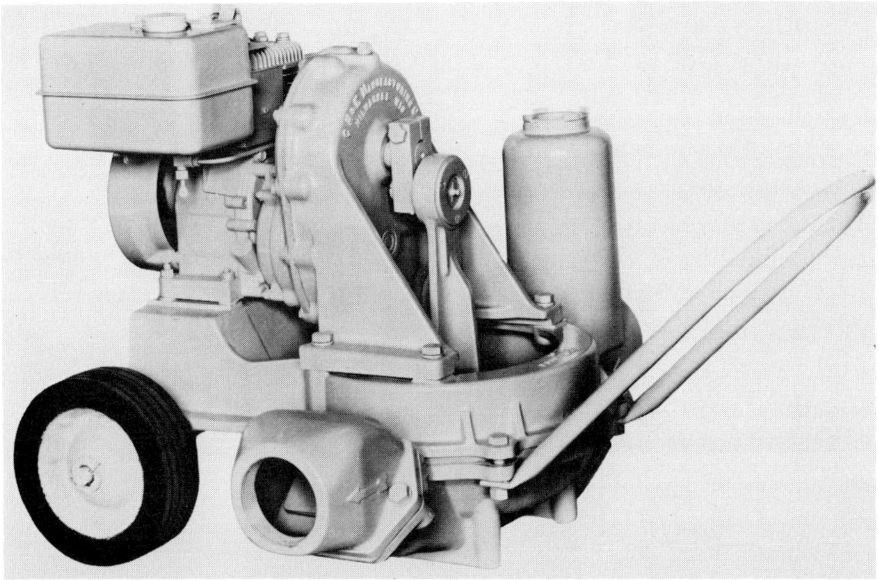

Figure 5-6 Diaphragm type pumps can be used in debris-laden water without damage.

Figure 5-7 Portable pumps are often used for removing water, but can be susceptible to damage. However, they are less costly to repair, if damaged, than a fire apparatus pump.

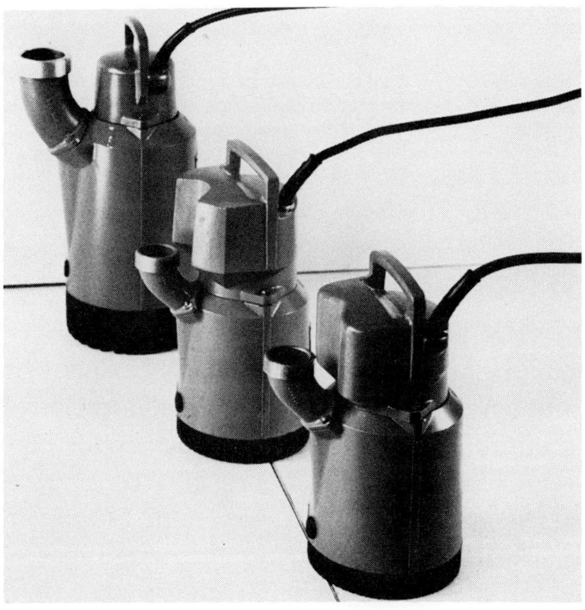

Figure 5-8 Submersible pumps are finding greater use in water removal. They can be quickly placed in operation to drain catch basins and catchalls.

Small quantities of water can be removed from floors in any area by using a portable water vacuum. This equipment operates in a manner similar to the ordinary vacuum cleaner except that the electric motor is arranged to enable the water which is collected to be stored in a metal or plastic container. The water vacuum is limited in capacity to the volume of the storage container. One typical type of portable water vacuum is shown in Figure 5-9.

Another variation to the salvage clean-up problem is a backpack type of water vacuum. The device is illustrated in Figure 5-10. Its use reduces water damage and simplifies clean-up operations. This type of water vacuum is completely portable and is very effective on all floor surfaces and coverings. These devices not only pick up water, but they also remove ashes, dirt and some fire debris. The water and debris that is removed from the floor can be trapped in the tank or discharged through an attached outlet hose through a window, door or drain. Care must be taken not to discharge debris into small drains.

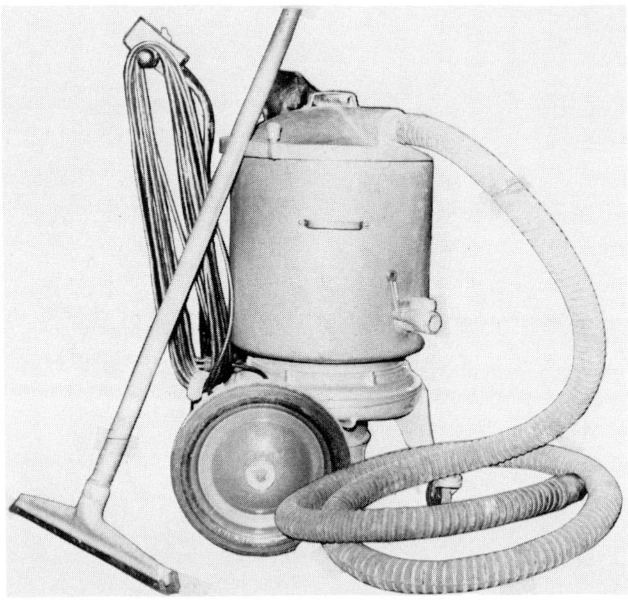

Figure 5-9 An industrial type water vacuum can be used during salvage to remove smaller quantities of water.

Figure 5-10 The backpack type of water vacuum is widely used by fire departments. It is efficient, but due to its limited capacity, must be emptied frequently.

OTHER WATER-MOVING DEVICES

When water collects in a building, most of it will obviously be on the floor. This fact alone indicates the need for consideration of the effects of water accumulation on floors or floor coverings. Although damage to the finish of wood floors is a concern, a greater concern is that wood floors react to severe or prolonged dampness by warping, swelling and buckling. Various types of composition and tile coverings will often separate from the floor when water seeps under the coverings. Carpeting quickly absorbs water and unless it can be dried, it should be taken up so that it and the floor can be salvaged. Professional maintenance and salvage firms in some cities can be called on a 24-hour basis for the specialized work of drying carpets. Additional tools and equipment used to remove excess water from buildings include brooms, squeegees, buckets, mops, shovels and sawdust.

Brooms and Squeegees

Brooms and squeegees are used to sweep or push water to doorways, scuppers, stairs or other points of drainage. Care must be taken, when moving water in this manner, to prevent damage to buildings and contents. Where necessary, objects that may be damaged should be protected with salvage covers or removed from the area of operation. Brooms are also used to sweep debris during overhaul operations.

Buckets, Mops and Shovels

Buckets are used to catch water dripping from upper floors and ceilings, to bail catch basins, and to carry water and debris from buildings. Some departments use standard custodian mop buckets with attached wringers to facilitate mopping operations. Mops are used to absorb water and to clean and dry floors, and scoop shovels are used to pick up water, sawdust and debris.

Sawdust

Sawdust is used to absorb water and to form dams and dikes. Some departments purchase sawdust in paper bags and carry a number of these on ladder and salvage trucks. Sawdust dams and dikes may be used to control flowing water or to direct it to a drainage point. Sawdust is also used to form seals between salvage covers and the floor.

REMOVING WATER FROM BASEMENTS

Since water will seek the lowest level, it is not uncommon in large fires for basements to fill with water. If several large streams are in use, the quantity coming in will quickly overpower the normal domestic drains which also have a tendency to become blocked.

Whenever possible, without causing undue exposure to their safety, firefighters should be assigned to limit water damage in a basement. This initially involves limiting or directing the flow of water by the use of salvage covers for dams, dikes or channels. If possible, basement drains, sumps or soil pipes should be used for water removal. These may not have sufficient capacity to handle the quantity coming in. In using any of these, adequate protection needs to be provided to keep debris from plugging the piping.

When internal devices are not sufficient, the fire department will have to develop other means of removing the water. Sump pumps and floor drains are often blocked during a fire. Those methods most generally used are portable pumps, submersible pumps (Figure 5-11), or dewatering devices. Dewatering devices are commonly called siphons or eductors. Many departments have these devices or can easily construct one (Figure 5-12).

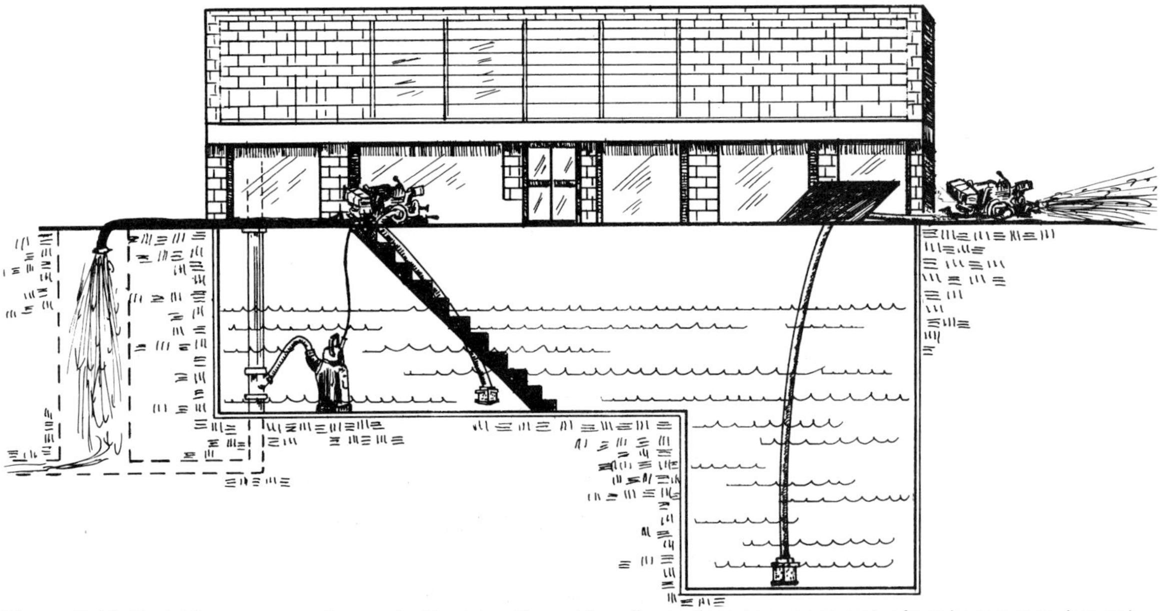

Figure 5-11 Portable pumps can be used with stairwells or sidewalk openings to remove water from basements. It may be beneficial to deliver the discharge into a sewer. Submersibles may also be used and can be discharged directly into soil pipes.

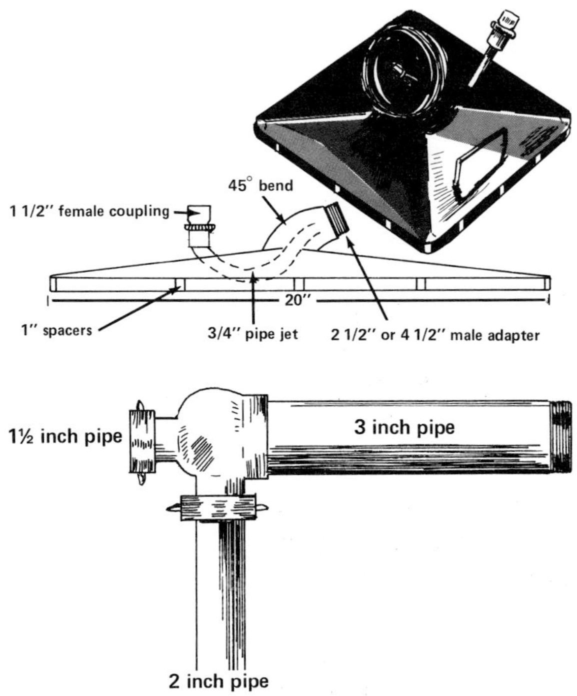

Figure 5-12 Dewatering devices can be used to remove dirty water from basements. They can be purchased or easily constructed and work on the basic venturi principle.

The dewatering device works by the principle of eduction similar to a highlift pump or foam eductor rather than by pumping. This eliminates any dirt or debris from getting into a pump. Supplying these devices with a 1½-inch hose can create a sizeable discharge. Larger lines can also be used to increase the flows (Figure 5-13). Pumps must be equipped with strainers and suction should be taken from the lowest point available. It may be advisable to discharge the water from the pumps into a soil pipe rather than run excessively long hoselines. If the water is discharged outside the building, it should be directed so it does not run back into the fire building or an adjacent building creating more water damage.

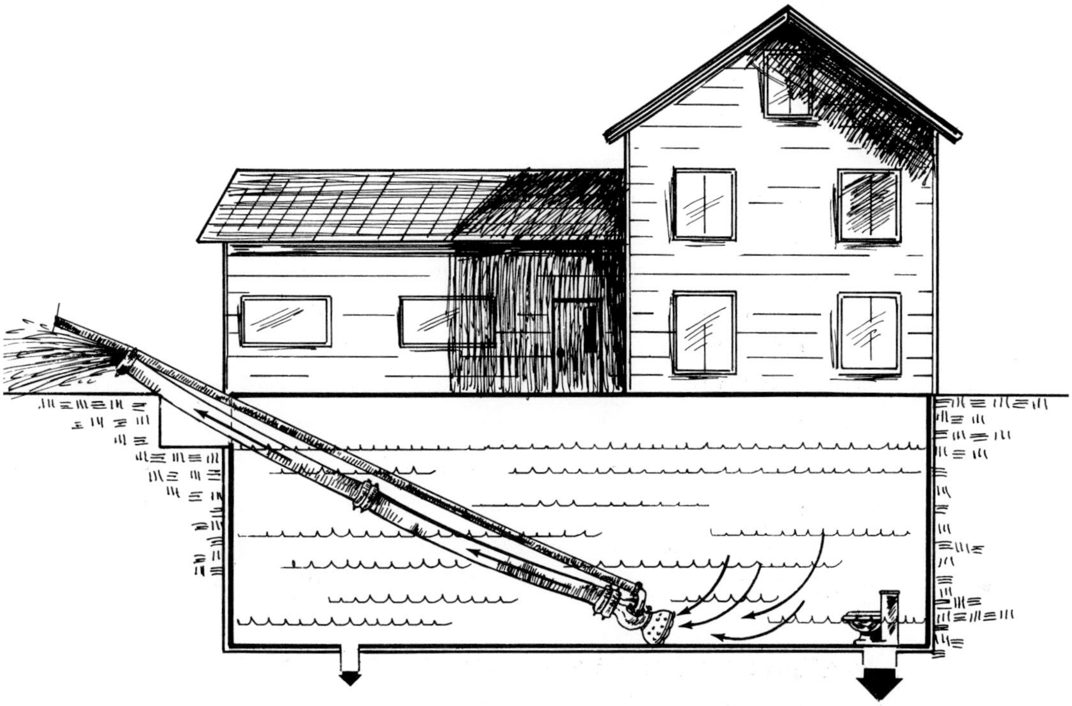

Figure 5-13 Large flows can be developed with 2½-inch intakes and suction hose used for discharge. Note that basement drains and toilets can also be used for water removal.

Checks must be made of adjacent buildings where party walls are involved for seepage or leaks. Salvage operations may be necessary in the exposed basement to protect stock. Additional manpower and equipment may be required if large amounts of water are penetrating the wall.

A major concern while removing water from basements has to be with safety. Before ordering men into a basement the possibility of floor collapse must be surveyed. Also the utilities must be checked especially gas and electric. If gasoline-powered portable pumps are in use, their exhaust must be directed to the outside to avoid a buildup of carbon monoxide. If possible, personnel from the building should be contacted before anyone enters to obtain information on any specific hazards in the basement such as holes, open tanks, hazardous storage or elevator pits.

Facilities in basements should be restored

An important part of the salvage operation in a basement has to do with restoring vital facilities after the water has been removed. Although this work is not the direct responsibility of the fire department, the owner should be given as much cooperation as possible under the circumstances. Included in this are having utility companies restore electrical power and natural gas for furnaces and hot water heaters, placing any refrigeration equipment back in service, and seeing that electric motors for vital operations such as ventilating systems or elevators are restored. The absence of heat during cold weather can cause considerable additional damage. In addition, the lack of heat will delay the drying of the building and could increase mildew and warping.

USING HOSELINES FOR ROUTING WATER

An extended length of booster line may be used as a temporary dike or a long continuous squeegee for clearing an area of debris and water. The booster line is outlined around the walls or perimeter of a room or area to be drained of surface or overflow water. By pulling the looped hoseline water can be moved toward the exit.

Larger hoselines may be used as dikes for routing water when the number of covers is limited or when there are no covers at all. Two lines can be used parallel forming a chute, or used singly as a dike. Hoselines can also be used as a catch basin.

WATER CONTROL FROM SPRINKLER HEADS

Understanding sprinkler head construction will aid water control

Standard sprinkler heads are designed to hold back water or air under pressure by a plug and a series of levers with a fusible link, or a quartz bulb. The functional parts of a soldered link sprinkler head and how they operate are shown in Figure 5-14. Another type of sprinkler head is the frangible bulb operating element. A small, specially designed glass bulb partially filled with a liquid holds the plug in place. When heat expands the liquid, the air bubble is compressed and absorbed by the liquid. As soon as the bubble disappears, the pressure rise within the tube is rapid. The bulb then shatters to release the opening cap or plug. The size of the bubble regulates the operating pressure and temperature. Typical frangible bulb sprinkler heads are illustrated in Figure 5-15. Sprinkler heads may be installed so that they are either pendent (point down) or upright (point up). Pendent installations are generally used so that the supply piping can be concealed above the ceiling. Upright installations are usually for dry-pipe systems so that water will not be trapped in

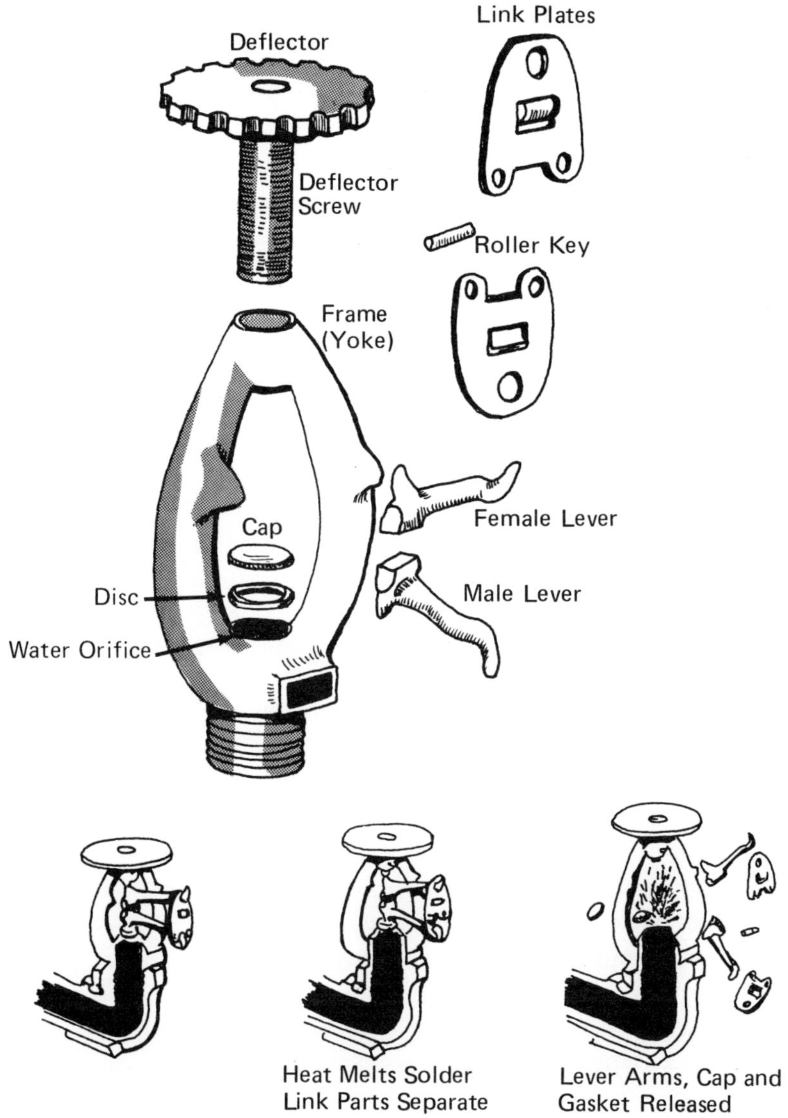

Figure 5-14 A fusible link sprinkler operation and the major components are shown to illustrate where stoppage of water flow will be necessary.

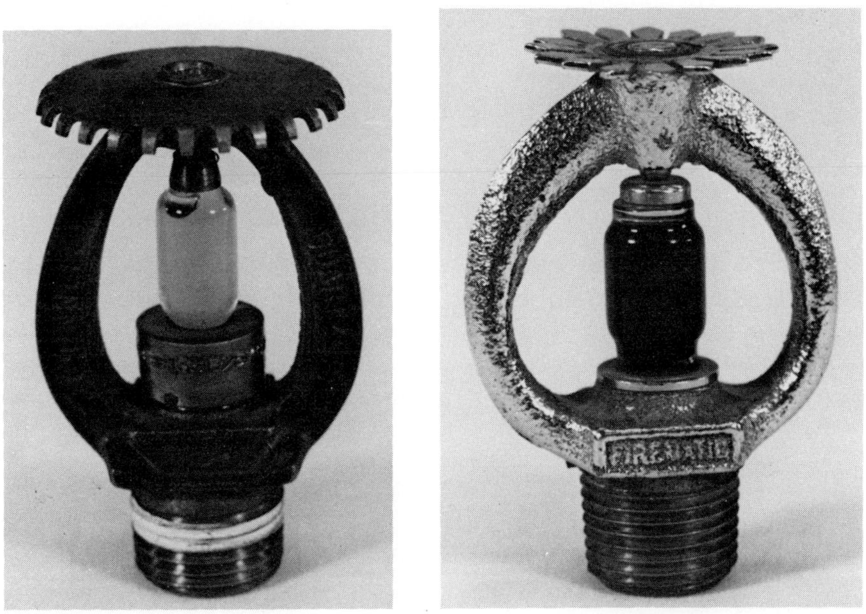

Figure 5-15 Automatic sprinklers may use a frangible bulb operating element. Here the liquid contained inside the bulb expands when heated bursting the bulb and releasing the plug.

the drop pipe and freeze. When pendent installations are necessary on dry-pipe systems, a specially designed "dry pendent" sprinkler head and drop pipe is used. Pendent installations also provide for a flush type sprinkler head as illustrated in Figure 5-16. There are several other types of sprinkler heads in use which are beyond the scope of this manual. For complete information on sprinkler systems reference IFSTA 210, **Private Fire Protection and Detection Systems.** (Not available as of Jan. 1979, write for details.)

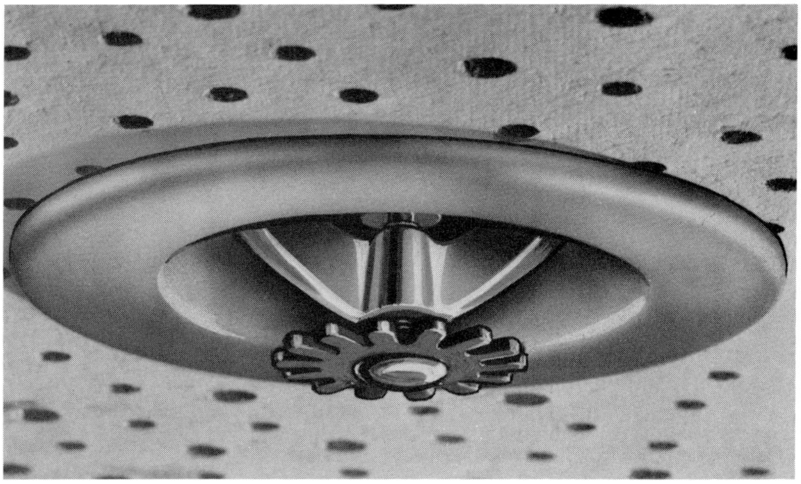

Figure 5-16 A flush mounted pendent sprinkler head is used for a decorative affect in many installations.

Considerable time may elapse from the activation of a sprinkler head until authorization to close the main sprinkler control valve is received. Extensive water damage can result unless a flowing sprinkler head is plugged when it is no longer needed for fire extinguishment, but before the main valve can be closed. Obviously, a firefighter will get wet during the operation and should wear full protective clothing. Fire department ladders should be used to reach the flowing sprinkler head. There is no set rule for positioning the ladder, but in either case another firefighter should hold the ladder in position. Most wedges are so designed that they can be driven into position with the heel of the hand. If the wedge is properly made, practically all of the water flow can be controlled by the driven wedge. Wedges can have a band of rubber placed on them to improve their operation. Sprinkler tongs are, to some extent, more positive in their ability to hold back water. If properly applied, the rubber or neoprene stopper will permit no dripping from a secured sprinkler head. The procedure for applying sprinkler tongs into a flowing sprinkler head is illustrated in Figure 5-17.

Figure 5-17 Sprinkler tongs can be applied to stop the flow of water and enhance salvage operations without reducing the fire protection in other areas.

Sprinkler control valves should never be closed unless the fire is definitely under control. When the sprinkler system must be shut down to control water flow, the firefighter already detailed to the appropriate valve is notified. Floor or sectional valves should be used to control the flow if possible. Immediately after the correct valve has been closed, drains (main or auxiliary) must be opened. This allows water remaining in the system to flow harmlessly to the exterior of the building rather than into the building through open heads. The firefighter assigned to close a sprinkler valve must remain at the valve so that it may be quickly reopened if necessary.

An outside screw and yoke valve may be installed in a valve pit, on a main riser or on auxiliary lines. The main control valve for the entire system is an OS&Y installed in a pit or at the base of the main riser. OS&Y valves on auxiliary lines may be used to control water flow to various floors or building sections. As its name implies, this valve has an outside screw indicating the position of the gate. When the screw is extended beyond the valve wheel, the gate is open (Figure 5-18).

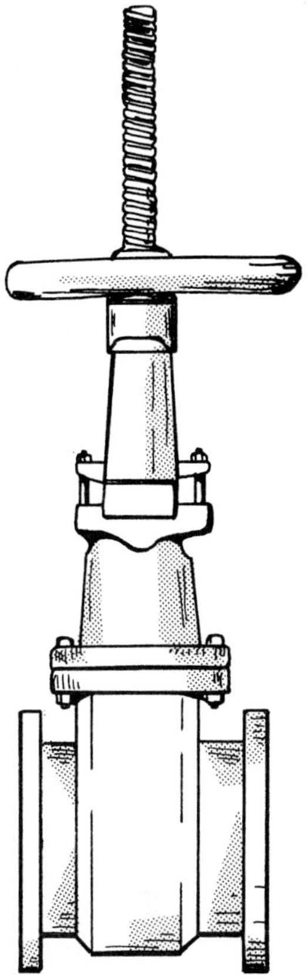

Figure 5-18 This outside screw and yoke valve is shown in the open position. Having taken adequate precautions, valves of this type may need to be closed for water control.

The post indicator valve is located outside of the building and has a sign that indicates whether the gate is open or closed (Figure 5-19). The valve also has a wrench usually sealed to a hasp on the post. In operating the valve, the seal must be broken and the wrench removed, turned over and placed on the valve stem. An adjustable hydrant wrench may be used if the post wrench is missing. Although the post wrench should not be padlocked to the hasp, this is sometimes the case. Firefighters should not hesitate to break such a lock if the valve must be closed to shut down the system limiting water damage.

Figure 5-19 A locked wrench and post indicator valve is shown. When salvage operations necessitate, the lock should be cut and the valve closed.

A variety of tools can be used to control and move water. Forming channels with hoselines or sawdust will limit the lateral spread of water being removed.

6

Overhaul Procedures

Fire departments have practiced overhaul operations since the first fire companies were organized in an effort to do an efficient job. Although these operations were, in essence, overhaul; they were not always recognized as such. The equipment that may have been used was normal auxiliary fire fighting equipment, and the search for and extinguishment of hidden fires were accepted as a part of the fire department's function. Determining the cause of the fire and recognizing and preserving evidence of arson were not, at that time, considered to be a fire department function.

As fire departments assumed the responsibility of salvage operations, it became evident that leaving the building in a safe condition was also a responsibility that could not be ignored. Fires of intentional origin have caused added responsibilities for the fire service and these responsibilities have brought about a request to recognize overhaul as a separate function.

SAFETY OF FIREFIGHTERS AND OTHERS

The primary objective of overhaul has long been considered to be the extinguishment of remaining small, hidden fires with a minimum of damage and the prevention of rekindle. Actually the first step in overhaul should be to provide safety in the fire area. Hazards to personnel during overhaul are present in that firefighters may be injured from falling through a hole or weakened structural area, by being struck by falling objects or tools, tripping, or stepping on dangerous objects. They are susceptible to cuts, puncture wounds, eye injuries, strains, sprains and broken bones during overhaul procedures. The toxic fumes and contaminated runoff a firefighter may encounter can also be harmful. Once a fire has been brought under control, time is available to plan and organize the overhaul phase in such a way as to provide the highest possible degree of safety. Overhaul should normally start at the top and work down in the structure. The steps required to establish safe conditions include: (1) inspecting the premises; (2) developing a sound operational plan; (3) providing suitable equipment; and (4) eliminating hazards. Overhaul safety will be enhanced by thinking during the use of hands and tools.

Overhaul starts at the top

Injuries during overhaul from exposure or toxic fumes have been on the increase. These injuries may be the result of hazardous chemicals, insulation decomposition, pesticides or polyvinylchloride (PVC) deterioration. Firefighters may come in contact with the products by inhalation, ingestion, skin or eye absorption. Since such unsuspecting items as flameproofed cellulose insulation can give off harmful boric acid dust, wearing proper full protective clothing including positive-pressure self-contained breathing apparatus is necessary to alleviate the many hazards during overhaul. All firefighters should be trained to recognize the symptoms of poisoning by these products, the immediate first aid measures and how to obtain medical assistance. It is important to advise attending physicians of the exact name of the chemical or pesticide involved. An uncontaminated label, if available, should be supplied to the hospital or poison control center. Exposure to contaminated runoff can also be extremely dangerous. Overhaul may have to include control of runoff, neutralization and absorption with proper disposal. Firefighters performing overhaul may have to be decontaminated before leaving the area and must take adequate care before smoking, eating or drinking. When necessary, technical assistance should be sought through Chemtrec (1-800-424-9300) which has access to manufacturers and to the Pesticide Safety Team Network. A great deal of the chemical hazards encountered in overhaul can be anticipated and proper precautions taken if preplanning has been completed for the location. For further information on hazardous chemicals and pesticides reference IFSTA 305, **Hazardous Materials.** (Not available as of Jan. 1979, write for details.)

Provide for safety in the overhaul area

Firefighters who were on the first-arriving units should be allowed to return to the station if at all possible. Exhausted firefighters may be more susceptible to injury. Later-arriving units or relief personnel should complete overhaul operations if they are available.

The foregoing purpose of overhaul is primarily directed toward providing safety for firefighters. However, other persons may also be on the premises such as representatives of the press, public utility company personnel, insurance agents and adjusters, investigators, and owners or occupants. The fire department is obligated to protect them as well as its own personnel. Providing safety for all persons who must have access to the fire premises should be the goal of this phase of overhaul.

VALUE OF PROPER OVERHAUL

The application of proper overhaul techniques provides valuable results. These results are important directly to the fire service and indirectly to the public. Although the fire service obviously profits most from proper overhaul practices, there is a hidden value to the public which is often realized through conviction of arsonists, reduced insurance costs and a quicker return to production schedules. Some of the values of overhaul are as follows:

Helps Locate Hidden Fires

Benefits of overhaul

Proper overhaul involves a systematic handling of contents, merchandise and debris to accomplish many purposes. One of these purposes is locating hidden fires so that they may be extinguished, thus preventing further damage. Hidden fires can be detected by touch, smell or sight.

Helps Prevent Rekindle

Proper overhaul does not assure that a fire will not rekindle, but it greatly reduces the possibility. Its practice enables firefighters to detect those materials which are more likely to rekindle, remove them from the premises and thoroughly soak them. Where it is unsafe to fully dig through the rubble, or debris is of such quantities that mechanical devices will be required, wetdown lines and crews should be put in service to control any flare-ups and avoid a rekindle.

Helps Determine Cause of Fire

Proper overhaul will aid in determining the cause of a fire. A firefighter's ability to recognize the cause is another problem. A fire cause involves two conditions, the fuel supply and the heat source. A fire results when a "heat source" hazard and a "fuel supply" hazard are permitted to come together in the presence of oxygen. During overhaul both the fuel involved and the heat source that caused the ignition should be sought to determine the cause of the fire. Only by including both parts of a fire cause in all reports can really meaningful statistics be compiled for use by fire prevention personnel.

Aids Recognizing and Preserving Evidence of Arson

Determining the cause of the fire goes hand-in-hand with recognizing and preserving evidence of arson. Proper overhaul will permit critical evidence to remain untouched and undisturbed, and it will also permit the safeguarding of other evidence that cannot be left at the scene of a fire. All firefighters should be familiar with the latest legal rulings on inspections and gathering information for the cause of a fire.

Helps Prevent Unnecessary Damage

Overhaul aids the fire service and the public

Successful overhaul should follow a step-by-step procedure or pattern since a systematic approach will be beneficial. Unnecessary damage to the building and its contents can be avoided by removing heavy water-soaked materials immediately. In many cases items can be removed from the building for final extinguishment such as is commonly done with mattresses and stuffed furniture. Overhaul can proceed from top floors down, from the interior rooms outward, or from a number of entrances simultaneously to eliminate additional damage as debris is hauled through an area. Restoring heat, covering open windows or roofs, and draining flush boxes or water pipes in unheated areas are all overhaul functions which help prevent unnecessary damage.

Permits Removal Of Original Fire Cause

A routine system of overhaul not only helps determine the cause of the fire, but it pinpoints the origin and permits the removal of items which could cause recurrence. A recurrence can best be explained by example. Suppose the cause was determined to be a faulty extension cord which ignited a nearby curtain. This finding affords an opportunity to remove the faulty cord prohibiting another fire.

Aids In Placing The Building In Safe Condition

Numerous unsafe conditions may exist in and around the fire building after the fire has been controlled. In order to work inside or near the structure it may be necessary to correct these. In some cases, dangerous conditions will not permit firefighters to enter the building and do a complete overhaul. In other cases, demolition may be the only feasible answer. The correction of unsafe conditions such as hazardous processes, damaged utilities, structural defects, damaged contents, unstable objects and open pits presents a challenge to every fire department.

Helps Improve Public Relations

Public relations is not a planned procedure by which personal gain is achieved, nor is it a "packaged service" obtainable for a price. The price that is paid by fire departments for good public relations is measured by their efforts to cooperate with property owners, law enforcement officers, utility companies, city officials and the public in general.

OVERHAUL EQUIPMENT

The salvage that is done before and after a fire fighting attack will have a direct bearing upon any overhaul that may come later, and many of the same tools and equipment are used for other operations. Overhaul equipment, tools and accessories generally consist of the normal complement of fire department tools, such as forcible entry tools, small hoselines, ropes, lights and portable extinguishers. In addition to these various fire department tools, some overhaul equipment may also be segregated and carried in overhaul kits. Unlike salvage equipment, the number of overhaul kits is limited to arson; safety; door, window or roof; and possibly another for some localized purpose. Overhaul kits should supply the selected materials necessary to effectively perform a specific function. When overhaul kits are employed, it is not necessary to hunt for supplies when a particular job needs to be done. Three suggested overhaul kits are as follows:

Overhaul tools may be assembled in kits

Arson Kit

An arson kit is to be used for collecting, protecting and preserving evidence of intentional fires and to help in determining the cause of the fire. The following suggested equipment is recommended for a basic kit:

- Squares of glass
- Assortment of metal containers
- Magnifying glass
- Assorted large-mouth jars
- Plastic bags
- Sealing tape
- Bulb syringe
- Flash camera
- Envelopes
- Scrapers
- Flashlight
- Bottles
- Labels

Safety Kit

This should contain tools and equipment that are needed to provide adequate warning to the public while firefighters are making a building or an area safe. The following equipment is recommended for a basic kit:

- Warning lights
- Carbon monoxide detector
- Combustible gas indicator
- Normal and chemical flares
- Signs and notices
- Explosion-proof flashlight
- Small rope

Door, Window Or Roof Repair Kit

This kit is to be used for the temporary repair of doors, windows or holes in roofs. The following suggested equipment for this kit is recommended:

Assortment of roofing and masonry nails
Heavy-duty stapler and staples
Adequate supply of roofing paper
One hammer hatchet
One claw hammer
One linoleum knife
Small bundle of wood lath
Roll of polyethylene plastic

DEBRIS HANDLING EQUIPMENT

Bulky objects susceptible to deep-seated burning, such as upholstered furniture, chests of clothing or mattresses must usually be opened, and hot particles may have to be completely soaked or submerged in water. In addition, materials of possible value should be recovered from the debris. Care must be taken that materials susceptible to smoldering or spontaneous heating are not covered with other materials.

Debris handling is the safe and efficient transportation of materials

A safe, efficient method must be provided for transporting materials to an overhaul location. The method of transporting can vary from hand carrying to the use of loaders and other mechanical devices. Debris may be placed in carryalls, empty containers, cans, or old salvage covers to protect the areas through which it is transported. Valuable articles should be removed from the path of transport, and other items should be protected with floor runners and salvage covers.

Pike poles, rubbish hooks, forks, shovels, hay hooks and similar standard tools are readily adaptable to debris handling operations. Two pieces of equipment especially designed for debris handling are carryalls and metal baskets. Mattress chains can be used to remove bedding and similar materials that can be folded.

Carryall

A small waterproof carrier or bag, called a carryall, that can be made from an old or damaged salvage cover is illustrated in Figure 6-1. This carrier is approximately six feet or eight feet square, depending upon the needs of the department. For carrying purposes, handholds are provided by lashing a sash cord through eyelets which are reinforced by grommets around the edges. The handholds permit the carrier to be puckered, thus

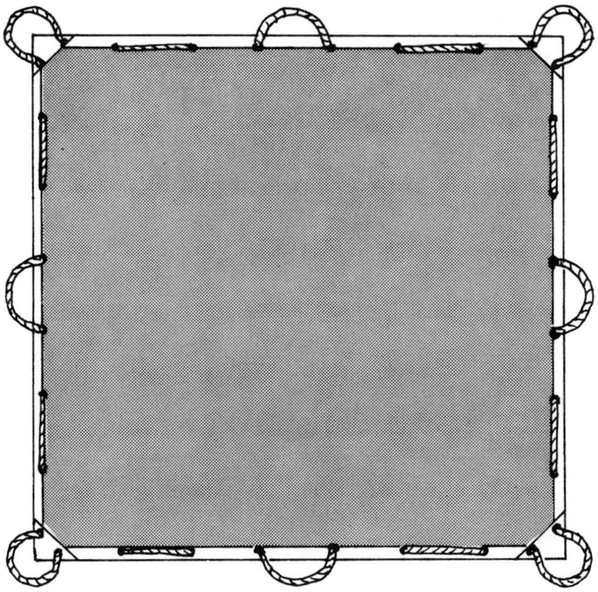

Figure 6-1 A carryall is useful in overhaul work to carry debris or to provide a watertight basin for immersing smoldering materials.

forming a bag. Such a carryall is useful in carrying debris, catching debris and providing a water sump for immersing small burning or smoldering objects when they cannot be easily removed from the fire building.

Making a carryall six feet or eight feet square has the added advantage of permitting its use as a mattress carrier (Figure 6-2). When used for this purpose, it should be wetted down before a burning mattress is loaded. A safer method is to insert an asbestos blanket inside the carrier which will reduce the possibility of burning holes in the canvas. A mattress can then be carried out of the building without the danger of scattering smoldering cotton

Figure 6-2 A carryall can also be used in many instances to remove a mattress.

or cloth about the premises (Figure 6-3). Mattresses are difficult to carry under ordinary conditions and especially if they are partially burned. Under these conditions the carryall is very useful (Figure 6-4).

Figure 6-3 By properly covering a smoldering mattress it can be removed without further spreading damage throughout the premises.

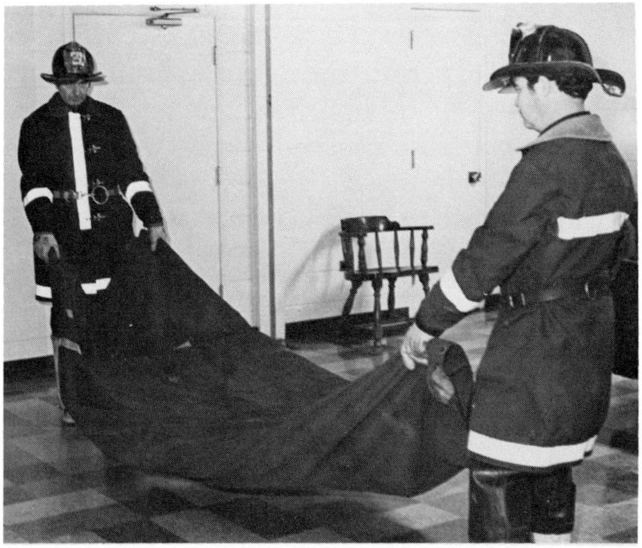

Figure 6-4 A mattress in a carryall can be carried outside and final overhaul and extinguishment accomplished.

Mattress Chains

Mattress chains help secure a rolled mattress for easier removal. A partially burned mattress can be difficult to handle without the correct equipment. In residential occupancies mattresses frequently must be removed to the outside for complete extinguishment. Mattress chains or straps are a great assistance in the removal process.

Mattress chains may be purchased from equipment dealers or made in departmental shops, usually of light chain or flexible cable. They are approximately eight feet in length and have hooks or similar locking devices to keep them tight around a rolled mattress. A mattress with a chain at each end is easier to handle than one wrapped with a single chain (Figure 6-5).

Figure 6-5 Mattress chains can be used for mattresses which will fold or roll holding the fire within until removed.

For the safety of firefighters during overhaul full protective clothing should be worn including positive-pressure self-contained breathing apparatus.

Subsequent Fire Control Operations

Subsequent fire control is simply a follow up after the initial extinguishment of a fire to prevent a rekindle. In this sense, overhaul procedures are functionally related to fire extinguishment. Subsequent fire control operations further entail the restoration of private protection devices and the premises in general. Other items which fall into this category are providing protection from the elements, making conditions safe and releasing the premises.

SEARCHING FOR HIDDEN FIRES

After assuring safety, one of the first routine operations will be searching for hidden fires. It may be true that sometimes a rekindle is due to inadequate overhaul, but there are also conditions where rekindle is due to circumstances beyond the control of the fire department. There are certain types of occupancies where complete extinguishment is extremely difficult, and a continued watch must be maintained over the fire area for a prolonged period.

Concealed Spaces

Methods of opening areas of buildings where fires may be concealed are covered in IFSTA 101, **Forcible Entry, Rope and Portable Extinguisher Practices.** It is the purpose here to discuss these concealed spaces from the standpoint of where to search.

Some of the more common concealed spaces in buildings are in walls or partitions, attics or cocklofts, above ceilings, below floors, in basements, air shafts, pipe chases and ducts.

Since it is extremely difficult to outline all of the operations that may be performed in searching for hidden fires, only the more fundamental conditions will be discussed. One prerequisite, before starting a search for hidden fires, is to determine the safeness of the building in the area to be searched. The degree of intensity which the fire attained during burning, and the amount of water used during its control are two important factors that alter the condition of the building. The first determines the extent to which the structural components have been weakened, and the second determines the additional weight on floors and walls because of unremoved water and the absorbent qualities of the stock. Consideration given to these two factors may prevent unnecessary loss of life during overhaul because of building collapse.

Consider building collapse before overhaul

An important objective, in searching for hidden fires, should be to make a systematic and careful check to determine whether the fire extended to other areas of the building or to other buildings. If it is found that the fire did extend to other areas, it is then necessary to determine through what medium it traveled. When floor beams have burned away at the ends where they enter a party wall, it is a good policy to carefully overhaul the ends of the beams. If there is any possibility of the fire extending, the ceiling on the opposite side of the wall must be pulled, if this was not done during the fire, to check the extension. Care must be taken to assure extinguishment without causing unnecessary water damage in the ceiling area beyond the wall.

Check all routes of fire spread

When the fire has burned around windows or doors, there is a possibility that there is fire remaining within the casings. To locate this fire requires the removal of the facings to assure complete extinguishment. Remove the top facing, then the side facings, remove the window frames, cut the sill at the center and pry it out, pry casing sides inward, and finally remove the entire casing. Another point of possible trouble is behind the cornice. When the fire has burned around a combustible roof or cornice, it is advisable to open the cornice and inspect for hidden fires. Two types of cornices are illustrated in Figure 7-1.

When concealed spaces below floors, above ceilings, or within walls and partitions must be opened during the search for hidden fires, the rug should be rolled and the furnishings of the room should be moved to locations where they will not be damaged. Only enough wall, ceiling or floor covering should be removed to assure complete extinguishment. Hidden fires in concealed spaces can often be detected by feeling with the hand, by smelling, by discoloration, blistering or by smoke. Checking around

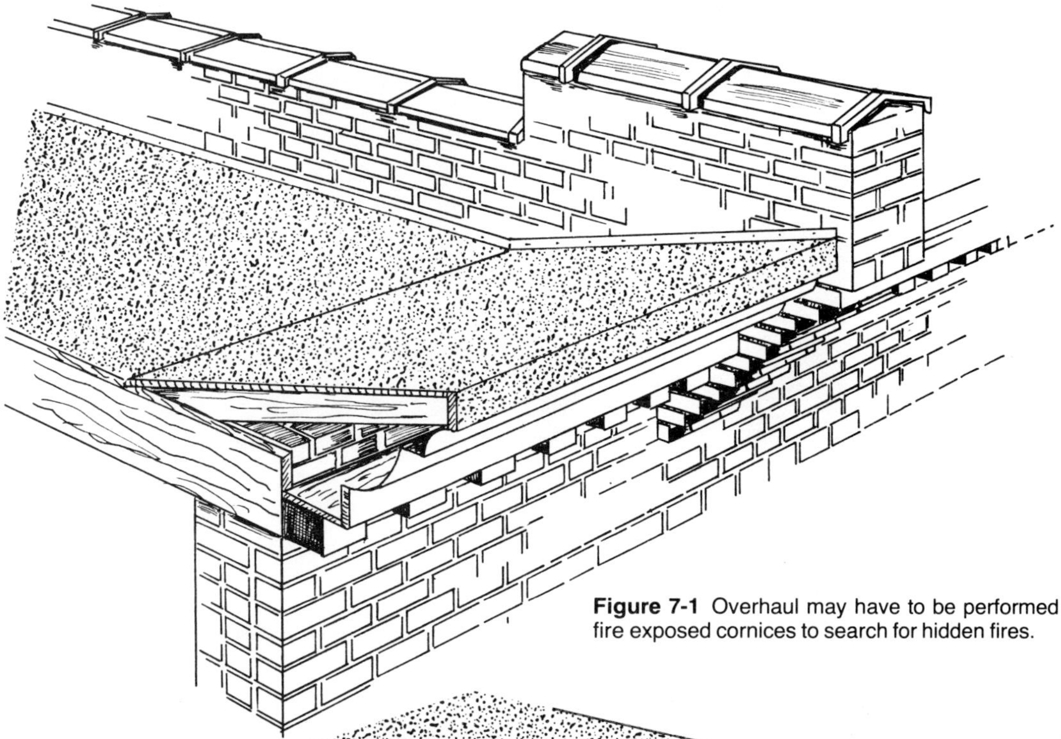

Figure 7-1 Overhaul may have to be performed by opening fire exposed cornices to search for hidden fires.

plumbing fixtures for extension into the wall should not be overlooked. Also, insulation material will often harbor hidden fires for a prolonged period, a condition which requires that these materials be removed in order to locate the hidden fire (Figure 7-2).

Figure 7-2 Overhaul encompasses a number of operations such as removing dangerous glass, searching for hidden fires around window casings, plumbing chases, or within walls, and constantly limiting damage with proper debris removal.

Fires in concealed spaces may be detected with devices recently adapted for use by the fire service which convert radiation energy in the form of heat into an electrical signal for display. More specifically, these infrared scanners operate on the amount of infrared energy which is radiated from objects based on their temperatures.

Infrared scanners are useful to the fire service in several ways ranging from locating the fire upon arrival to determining where hot spots are during overhaul. The unit cannot "view" through solid obstructions such as walls, doors and boxes because the scanner views the surface before it and compares the different temperatures across the surface. Those portions which are hotter than others will show up on the viewing screen as brighter red or black in one model. If, however, a fire is located behind a wall and has heated the wall, this will be evident as a red section in the wall surface. The image depicted on the screen will usually easily distinguish the object that is being viewed, i.e. a person, bricks, junction boxes or lighting fixtures. As the viewing range increases, the discernibility will diminish. Much damage has been averted by scanning the area to determine if there actually was a hot spot still present or just residual heat.

Under Debris and Around Merchandise

Too much emphasis cannot be placed upon the importance of searching for hidden fires under debris and around merchandise. Because of the absorbent qualities of materials such as paper, wood, fabrics and insulation that are likely to be in piles of debris, it is considered to be the best policy to remove the material from the building whenever possible. Not only are such materials a dangerous hazard because of the added weight and expansion, but it is frequently difficult to determine the exact location of the hidden fire within the pile. This condition is particularly true if the pile of debris is large or contains baled goods. Fallen structural members of the building often form voids in which hidden fires smolder and later rekindle.

EXTINGUISHING HIDDEN FIRES

It is very essential for firefighters to wear proper protective clothing including positive-pressure self-contained breathing apparatus while performing overhaul and extinguishing hidden fires. It is particularly important for metal insoles to be worn inside boots, and departments should require gloves and goggles to be worn as added precautions. This protection is an aid to the tedious and hazardous task of extinguishing hidden fires.

Provide charged hoselines during overhaul

Charged hoselines should always be available during the extinguishment of hidden fires although the same caliber of lines as were used to bring the fire under control are not always essential. Fire department pumpers can often be disconnected and returned to the fire station, but one or more 2½-inch hydrant lines should be left for standby. These mobile 2½-inch lines do not necessarily need to be tugged around by an overhaul crew during their search for hidden fires, but they should be kept ready to extinguish any major fire outbreak. Smaller 1½-inch hoselines are often extended from these 2½-inch lines to provide better maneuverability throughout the building. Booster hoselines are ideal for this type of service, but the use of booster lines involves having a pumper in operation during the overhaul period. Regardless of the size of the hose, the nozzle should be placed so that if accidentally opened it will not cause additional water damage. Whether or not a pumper should be employed should be determined by the officer in charge, based upon the need for such equipment and the availability of standby companies. Garden hose connected directly to a domestic water supply serves as a medium by which small, smoldering and hidden fires can be extinguished. When garden hose and other limited extinguishing devices are used during overhaul, provisions should be made to maintain large standby hoselines at a close proximity. Since a straight stream can scatter material, a limited fog stream is used during overhaul; however, a straight stream may be needed to bore into tightly packed material. Portable fire extinguishers provide limited extinguishing capability and are not as reliable as hoselines for overhaul purposes.

Quite frequently small burning objects are uncovered during overhaul. Because of their size and condition, it is better to dunk the entire object in a container of water than to try drenching it with a stream of water. Bathtubs, sinks, lavatories and washtubs can all be useful for this purpose. The previously described rubbish carrier (carryall) can be filled with water and used for this purpose. Portions of burning clothing, pillows, draperies and parts of upholstery may also be more easily handled by dunking them in water. Larger furnishings, such as mattresses, overstuffed furniture and bed clothes, should be removed to the outside where they can be easily and thoroughly extinguished (Figure 7-3). It is important for all firefighters to remember that all scorched or partially burned articles may prove helpful to an investigator in preparing an inventory or determining the cause of the fire.

Figure 7-3 Thorough overhaul can be accomplished outside the structure without additional water damage.

The use of wetting agents is of considerable value when extinguishing hidden fires. The penetrating qualities of wetting agents are helpful in complete extinguishment of hidden fires in cotton, upholstery, baled goods and a large number of other materials. If special fire apparatus that is designed to use wetting agents is available, the agent may be applied from a booster or 1½-inch hose. Many departments, however, carry a wetting agent in a container so that a portion can be added to water type extinguishers. The agent can then be applied when needed. Since dust can support a fire in a smoldering condition giving off little smoke, careful overhaul must be performed in dust areas to assure the fire is out. Wetting agents will be of assistance in these areas also.

OVERHAUL OPERATIONS AROUND DUCT SYSTEMS

Numerous duct systems are installed in buildings today ranging from the kitchen stove exhaust system to sophisticated ventilation systems in multi-storied buildings. These systems may be used for heating, cooling, removing dust or flammable vapors, disposing of waste or excess material and transferring products.

These duct networks, if not adequately protected, can extend the fire or associated damages. The construction of the systems adds to the difficulties since they are often inaccessible, poorly installed and have fans or blowers creating a forced draft. During overhaul all possible channels of fire spread must be checked for hidden fires.

Any leakage in the ducts which could spread heat or smoke to susceptible areas such as computer rooms must be stopped. Since there is the possibility of extremely large losses where computers are located, these installations should be indicated during preplanning and all necessary salvage operations performed to limit damage (Figure 7-4).

Figure 7-4 With adequate preplans complete salvage cover protection can be provided to computer installations.

Outside intakes in adjacent buildings may also need to be considered because smoke could be drawn in causing damage or creating a panic situation. A major problem is the common practice of many ducts terminating in an open attic or partition wall. This may require the employment of several overhaul techniques such as opening walls, roofs and possibly using ejectors to remove smoke to the outside.

Each duct system will require specific operations depending on its use and construction. The following general guidelines will bear consideration in buildings equipped with a duct system.

Consider duct systems from preplan to overhaul

- Determine the portion of the structure which contains ducts and their routing.
- If the system is operating, it should be shut down to alleviate any possible spread of heat, smoke, fire or fanning of hidden fires.
- All areas involved will need to be ventilated to the outside to allow effective examination of the duct system.
- Any false ceilings or framing enclosing the duct system must be checked during overhaul.
- Many duct systems are equipped with openings for cleaning which can be used in checking on fire spread or possibly to extinguish hidden fires.
- Water should not be used carelessly in or around the duct. Excessive weight could cause collapse of the duct or concealing members. Only apply water where there is actual fire, not to smoke moving in the duct.
- Duct system outlets must be checked for spread of fire to roofs, other buildings or areas downwind.

HANDLING DEBRIS

Non-salvageable material, such as lath, rags, paper and the like, may be scooped into a carrier, carried out of a building, and piled in a conspicuous place for future examination. Place the material in a safe place in case of rekindle. After the bulk of all debris has been removed, excess water on floors should be mopped or squeegeed dry. If the floor is wood, it is a good policy to take up excess moisture by covering the floor with sawdust for a period and then remove the sawdust. Damp sawdust should not be left on the floor longer than absolutely necessary, for it may damage the floor. During the removal of debris from buildings, careful observance should be made of any evidence of incendiarism that might be present. Should such conditions arouse suspicion, the area involved should be left intact until further investigation has been carried out.

DETERMINING CAUSE OF FIRE

During the thorough overhaul of a building, indications of the cause of the fire often will be found. Any information that will be of assistance to fire investigators should be noted. In addition to general facts that may be apparent, the area of the structure where the fire started and the point of origin may be discovered. Any significant conditions which may have encouraged the fire to start or possible ignition sources that are found should be preserved for the investigation team. Their analysis of the fire can be much easier if they do not have to completely sift through the debris. Several pair of eyes performing overhaul and at the same time alert for the cause will cover a large area in a shorter period of time than the single investigator. The immediate discovery of the cause may lead to the prevention of a similar fire within the community.

RECOGNIZING AND PRESERVING EVIDENCE OF ARSON

An official fire report should be completed for every incident. This report is usually made out by the officer or person in charge of the fire fighting operations. In making out his report the officer should try to determine the cause of the fire. If he feels that arson may have been involved, then he may want to call in an experienced investigator. For more detailed information on how to make out fire reports, refer to IFSTA 208 **Records and Reports For the Fire Service** and for further information on determining the cause of the fire refer to IFSTA 110 **Fire Prevention and Inspection Practices.**

Be alert to fire causes and unusual circumstances

Fires of intentional origin may be caused, responded to and extinguished without being investigated if the fire department personnel do not recognize the evidence that is sometimes quite pronounced during and immediately after fires. Those firefighters who arrive on the scene and make the initial attack, and those who perform the subsequent overhaul have the responsibility to watch for and recognize unusual or unnatural circumstances which possibly indicate a suspicious fire. Arson is a peculiar crime in that there are usually no complaining witnesses to the crime. Arson is often a result of planning and considerable preparation although arson committed on a spur of the moment decision is common. Since firefighters are normally not enforcement officers, trained arson investigators are summoned to investigate suspected individuals and conditions. Such inquiries might prevent placing firefighters or local authorities in a position of embarrassment.

The responsibility of a trained arson investigator is to aid in an investigation and prosecution of persons for arson and kindred crimes. His information must be gathered from circumstantial and physical evidence that is available at the time of the investigation. Recognizing and preserving evidence of arson is the responsibility of firefighters, without which a trained investigator will always be seriously handicapped. Local firefighters should become proficient in observing, recognizing and preserving evidence of arson, and organizing the evidence or information in a sequence that will benefit the investigator when he arrives. They must then communicate pertinent information to the investigator. The purpose here is to outline such activities for this use. This becomes even more important with recent court decisions requiring search warrants for re-entry into a building. There should be no discussion of the fire with outsiders if arson is suspected.

The following outline of activities for recognizing evidence of arson and the excerpt on protecting and preserving evidence of arson are abstracts from the National Fire Protection Association's publication, *The Firefighter's Responsibility In Arson Detection.* This publication was compiled by a subcommittee of the committee on Firemen's Training of the National Fire Protection Association. For a more complete study of fire investigation and preserving evidence reference IFSTA 604, **Fire Cause Investigation.** (Not available as of Jan. 1979, write for details.)

Protecting and Preserving Evidence of Fire Causes

Once discovered, evidence must be protected

Two things should be kept clearly in mind by firefighters regarding the protection and preservation of evidence of fire cause:

- Keeping the evidence where it is found, untouched and undisturbed, if at all possible.
- Properly identifying, removing and safeguarding such evidence that cannot be left at the scene of the fire as it was discovered.

No changes of any kind should be permitted in the evidence other than what is absolutely necessary in the extinguishment of the fire. Photographs are excellent supporting evidence if they are taken immediately. The photographs should be made before any evidence is washed away.

One precaution which must be exercised by all firefighters during the fighting of a fire is to avoid tramping over possible arson evidence and obliterating it so much that it becomes useless. The same precaution applied to the excessive use of water may avoid similar unsatisfactory results. Remember that overhaul and clean-up of areas which seem remote to the point of fire origin can destroy evidence vital to the investigation of the incident.

Protect and Guard Evidence

Evidence remaining at the scene of the fire can be protected in various ways, if and when discovered. Areas containing the evidence can be roped off, or goods and materials of various kinds can be piled around the evidence to keep people away until the arrival of the investigator. Guards can be posted to prevent tampering or needless handling of evidence. Prevent any unauthorized entry and even accompany the owner/occupant on his visit until after the property has been released by the fire department. Leave plenty of room around the evidence to protect it exactly as it was found. Protect several feet of discovered tire tracks, and if the car turned around, protect the print of all four wheels where the car turned. Detection work of this kind by firefighters makes it possible for trained investigators to later take measurements, photographs and make plaster casts.

The protection of several human footprints permits measurements of the prints, comparison of prints, estimate of length of stride, position of feet and any peculiarities in the gait (walk or run) of a suspect. It may also be possible to secure identifying marks on soles and heels of the shoes. Take similar precautions with animal footprints. Boxes placed over prints will prevent dust from blowing over an otherwise clear print and keep them in good condition for either photographs or plaster casts at a later time.

Preserve footprints, tire tracks, accelerants, containers and partially burned documents

Burned and partially burned papers found in a furnace, stove or fireplace should be protected by closing dampers and other openings immediately upon discovery. This operation will eliminate or reduce the draft. Wet rags may be used to smother the fire or to stuff stovepipes, but streams of water directed on such evidence should be avoided.

Call to the attention of the investigator any grease or oxidizer that is suspected to have had a part in the fire. Bottles which have been used to carry flammable liquids can sometimes be used to develop latent fingerprints, although this possibility is not general because of the smudging of the prints by the oil. It is, however, considered of sufficient importance to protect all such bottles from indiscriminate handling, so that if such prints do exist, they can be identified. Sometimes bottles filled with flammable liquids and tightly sealed are hung over the set fire so that, when they reach a sufficient temperature, the bottle will break and allow the liquid to increase the intensity of the fire.

Points On Protecting and Preserving Evidence

- Use care in the application of hose streams.
- Perform overhaul and salvage carefully.
- Have a witness to all discoveries.
- Record the time of day, date, location and circumstances under which evidence was discovered.

- Call the proper investigative authority to carry on with the evidence you have detected.
- Do not allow evidence to be walked over and trampled out.
- Install barricades around and properly guard evidence.
- Handle evidence as little as possible.
- Move evidence from the scene of the fire only when absolutely necessary.
- Have photographs taken at the scene of the fire when evidence must be removed.
- If persons are included in photographs, be certain they are identified.

Observing General Conditions at the Scene of the Fire

A. Upon approaching the fire:
   1. Observe the elements.
   2. Observe persons and automobiles.
   3. Observe color of smoke and flame.
   4. Observe size of fire and speed at which it is traveling.

B. Upon arrival at the fire:
   1. Observe the area of origin.
   2. Observe the number of separate fires, their intensity and rapidity of spread.
   3. Observe odors, methods and difficulties encountered in extinguishing the flames.
   4. Observe the source of ignition and material ignited.

Observing Conditions at the Time of the Fire

A. Conditions outside the building.

B. Condition of the building:
   1. Observe whether the doors and windows are locked.
   2. Observe the condition of doors, windows and locks.

C. Condition of owners, occupants and bystanders:
   1. Observe the dress and manner of persons at the fire. (Pump operators can sometimes scan the crowd or talk to bystanders.)
   2. Look for familiar faces.

D. Condition of contents:
   1. Watch for indications of the removal of personal articles.
   2. Watch for indications of the removal of valuable merchandise.

Locating Evidence of Fire Causes

A. Look for indications of arson intent:
   1. Multiple fires.

2. Odors.
3. Area of most severe burning.
4. Fire damage to ceiling and floor areas.
5. Undue wood charring.
6. Uneven wood burning.
7. Holes made in walls and floors before fire.
8. Holes burned in floors possibly by accelerants.
9. Underside of contents and structural elements.
10. Inoperative sprinklers and fire doors.
11. Tracks, footprints and fingerprints.
12. Condition of heating equipment.
13. Containers used for flammable liquids.
14. Residues of wax or paraffin.
15. Debris and salvage should be removed in layers.
16. Watch for the "V" burn pattern.

B. Look for arson materials and arson equipment:
   1. Trailers between fires.
   2. Candles.
   3. Matches.
   4. Chemicals.
   5. Flammable liquids.
   6. Rags, waste and paper.
   7. Gas stoves.
   8. Glass (especially magnifying glass).
   9. Bottles.
   10. Unusual gadgets and timing devices.
   11. Electrical equipment.

Possible Indications of Arson

- Look for the unusual and unnatural
- The presence of incendiary plants or materials
- Multiple origins, or uncommunicated fires
- Fire preceded by apparent preparation, such as the removal of contents or the absence of contents that are usual to a place of business or dwelling. Removal of personal items
- Fires destroying obsolete, unsaleable or unuseable contents
- Peculiar conditions and locations of burning, such as in the center of the building, or unnatural fire spread
- Rearrangement of furniture or furniture used to block means of entry
- Closet fires
- Water heater fires
- The presence of a suspicious location of electrical appliances
- Fires which gave evidence of damaged fuel lines or electric wiring prior to fire.

- Open gas jets
- Paper or newspapers spread around
- Fires where damaged or pre-burned contents are found
- Fires on property listed for sale or during renovations
- Fires to cover up other crimes such as murder, theft or burglary

PREVENTING UNNECESSARY DAMAGE

Damage which would be sustained by a building after the fire can be curtailed by overhaul. Smoldering fires can be doused limiting further smoke damage, hidden fires are extinguished removing the chance of rekindle, water and debris are removed limiting the absorption of moisture and dirt, and many safety hazards are removed that could cause injuries. Adequate ventilation and possibly deodorization incorporated during overhaul will reduce further odor damage. If heat cannot be restored, the closing of water valves and the draining of pipes and fixtures during cold weather will prohibit additional water damage. This includes steam pipes, liquid tanks, sprinkler or product lines, toilets and elbows in drains among others which could freeze and burst. The drain of a bathtub should be plugged or protected before using the tub for final extinguishment of bedding, pillows and stuffed cushions to avoid plumbing damage. When opening a wall, first place a salvage cover on the floor to catch the debris, remove the baseboard and place an investigation hole behind the baseboard where it can be easily covered or repaired if no fire is evident. When fire is evident, remove the wall covering up as far as is necessary. Care should be taken to only remove as much wallboard, lath and plaster or metal ceiling as is necessary. If heavy materials are on upper floors and can be moved, they should be placed near the outside walls or over supporting columns or bearing walls to alleviate the possibility of floor collapse. An important item to preserve to assist in the determination of the loss is any water soaked or partially burned records or inventory documentation. When overhauling mercantile occupancies, packages or cartons should not be opened unless they have been in actual contact with the fire. Only destroyed items of no value should be placed outside or in the street. A great deal of care should be taken not to spill bottles, cans, pails or drums of materials. Additional damage can be caused by these goods mixing or reacting.

Overhaul reduces post-fire losses

A simple way to limit damage in dwelling fires is to limit the number of personnel inside on minor fires. This will keep dirt and material from being tracked around. A great deal of unnecessary damage can be prevented by simply considering the property as one's own and treating it accordingly.

REMOVING ORIGINAL FIRE CAUSE

In many fire situations after the overhaul operation has led to the determination of the cause, it is evident that the cause can be removed. The heat source or the fuel source may be removed in order to prevent recurrence of the fire. The safety of those performing overhaul may be enhanced by the removal of the fire cause. Often the occupants must be educated about the fire cause or they will commit the same mistakes again possibly endangering their lives and the lives of others.

Hidden fires must be sought during the overhaul operations. Concealed spaces where fire may travel such as around charred door frames should be opened and checked.

8

Restoring the Premises

One objective of salvage practices is to leave a building and its contents in as nearly the same condition when the fire department leaves as it was before the fire occurred. The building should be left in a condition that is as livable as possible. Overhaul work would not be complete if debris and water were left in a building for this practice would add to the damage already done by the fire. Partially burned articles should not be discarded, for considerable salvage may be involved, and sometimes parts of burned articles serve a useful purpose in making inventories of losses. The overhaul crew should not unnecessarily destroy articles or remove them from the scene until the origin of the fire has been determined.

MAKING THE BUILDING SAFE

Probably the first objective in restoring the premises is making a building safe. Hazardous manufacturing processes should be thoroughly checked by a competent technician to prevent additional damage by fire or explosion and to safeguard against the dangers of chemicals and gases. Utilities, such as natural or artificial gas lines and electrical service, should be shut off or disconnected as soon as possible. The supply lines or wiring should be checked for safety by a utility respresentative before services are restored. Another phase in making a building safe is checking its structural condition. Loose cornices, bulging or cracked walls, sagging floors, and weakened roofs should be removed or guarded with barricades to prevent needless injuries. Conveyors, hoists, escalators and elevators should be checked for safety before they are released to the property owner for public use. Trained technicians are usually available for this service and should be summoned to aid in this survey. All these, plus the added security of posting adequate guards about the premises, are precautions that should be taken if additional disastrous occurrences are to be held at a minimum.

Covering Roofs, Windows and Doors

Covering holes in roofs after fires is a very important operation for it is essentially a poor practice to salvage a stock of merchandise or household furnishings and later have them damaged by the elements. The type of roof and the size of opening determine the method of covering.

On a pitched roof there may be a small hole, a larger one or one near the peak. To cover a small hole on a pitched roof the roofing materials around the top of the hole should be raised with a claw, kelly or similar tool. The materials used to cover the hole are placed under the uplifted roofing to prevent water from running underneath. The sides and bottom of the cover are secured with roofing nails. If salvage covers are used, lath or boards should be used to hold down the edges as nails should only be driven in grommet holes. A hole near the peak of the roof would have the "roofer" placed over the hole and both sides of the ridge (Figure 8-1). Roofing paper could be used in a similar manner. The material used must be properly secured on all four sides. For larger openings requiring more than one cover, place the covers from the peak down lapping the top cover at least six inches over those placed below on the roof and secure the edges.

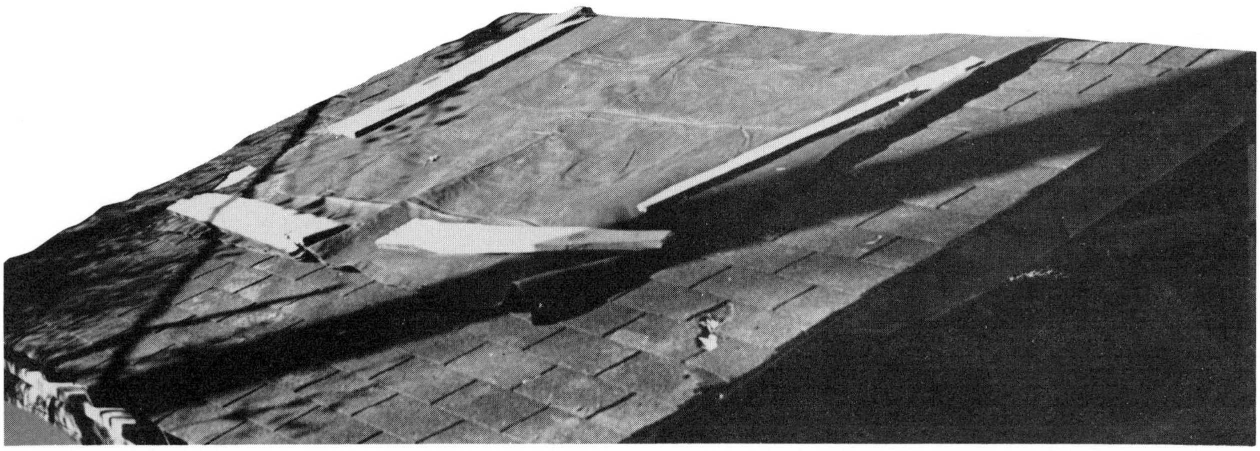

Figure 8-1 When covering holes near the peak of a pitched roof the cover should be placed over the hole and extended over the ridge before being secured.

When covering holes in a flat roof the edges around the hole are raised about six inches and blocked with available materials. For small holes a frame may be made from roof boards and secured around the hole. The frame raises the cover above the roof and provides for drainage. When covering a larger hole on a flat roof, the roofing paper or tin must also be raised and blocked with bricks or blocks of wood. A framework may be necessary to support the covering materials on a larger hole. This can be constructed by securing short wooden uprights at the edges of the hole and putting a support across the top. The cover is placed over this and secured around the edges by weighting or nailing (Figure 8-2).

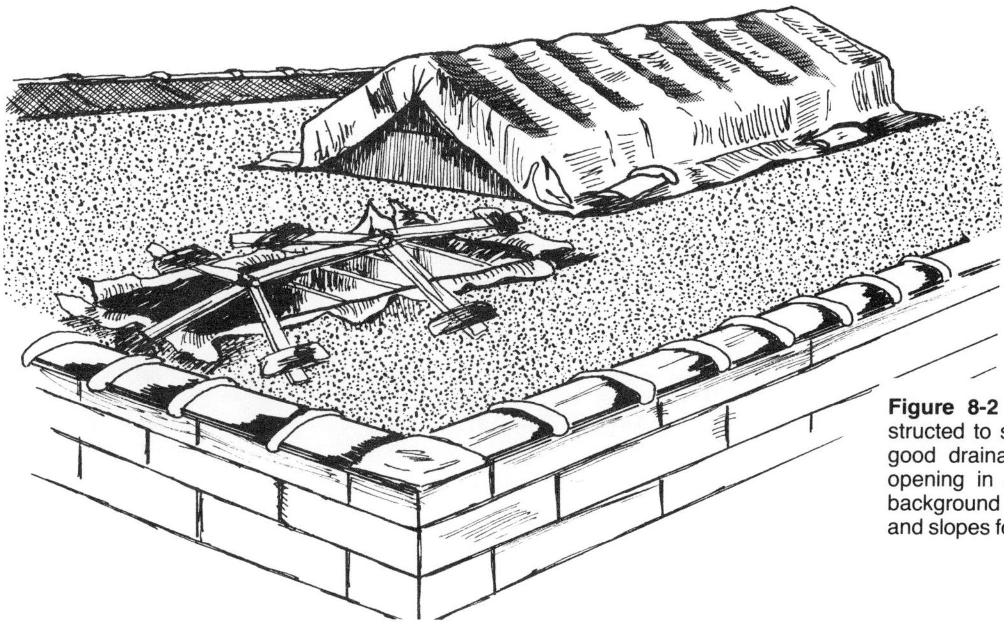

Figure 8-2 A framework can be constructed to support a cover and provide good drainage where there is a large opening in a roof. Skylights as in the background have their own framework and slopes for drainage.

In some instances it will be impossible to cover roof openings. When this condition exists the area below the roof is protected by "bagging." The ceiling joists or floor below the roof is covered forming bags or troughs between the structural members to hold any water (Figure 8-3).

Skylights, roof hatches and air shafts are probably the easiest of all roof openings to cover and make watertight because of the elevated framework (Figure 8-4).

Roof openings should be covered with roof covers, salvage covers, tarpaper or plastic sheets. The edges must be tied, weighted, or nailed down for security from the wind. Rolls of roofing paper and polyethylene sheeting can be used with staples and mastic to form watertight coverings. Many departments carry older salvage covers "roofers" for covering roofs; these should be appropriately marked. The size of the opening and length of time it will remain covered will determine the material the department uses on a roof.

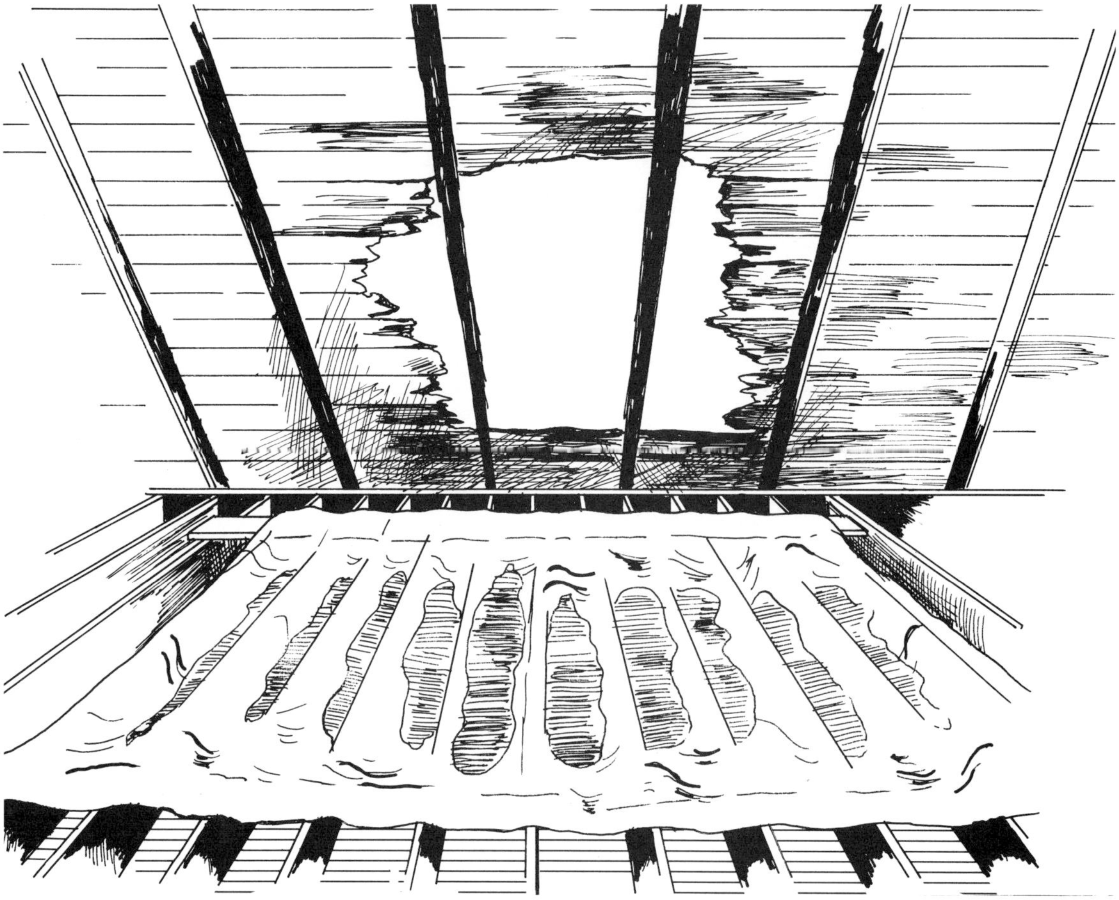

Figure 8-3 In those cases when it is not practical to cover a roof, it is possible to "bag" the area below and provide protection against the weather.

Figure 8-4 Skylights, due to their framework, are not difficult to cover and make watertight.

As with any aspect of fire fighting, extreme care must be taken while working on a roof. Since the roof has already been weakened by the fire below, before working on the roof test it for structural stability. Use extra caution when working on wet or ice covered roofs. Care must also be exercised in using covers on the roof so that they are not caught by the wind pulling a firefighter off the roof.

Cover all openings

Outside openings, damaged or removed windows or doors, must be covered during overhaul. A good grade of roofing paper makes a satisfactory cover for open doorways and windows. Roofing paper can also be held in place with strips of wood and nails. If the opening is in a masonry wall, masonry nails can be used to a good advantage. Since plate glass windows are larger, they will require a salvage cover to close. In order to keep the wind from blowing the cover and normal building breathing actions from stressing the cover, a brace will be needed behind the cover built from materials available at the site. Also secure the cover from the outside with a nailed board.

Making The Contents Safe

Be alert to content hazards

The damaged contents of a building can present serious safety hazards, and a careful check should be made to restore these situations to a safe condition. A serious situation concerns materials that have the ability to absorb large quantities of water. The absorbed water could add tons of weight to the already weakened floor structure, and it is often the best policy to remove the material from the building. This practice is especially important if weight conditions exist on upper floors. Although not specifically contents, several types of equipment and fixtures will also need to be checked. In moving stock or heavy equipment it may be necessary to use mechanical devices available at the location operated by plant personnel. These could include hydraulic hand trucks, lift trucks and overhead hoists. The following is a partial list of items which may be unstable or weakened and have to be moved or secured to enhance the safety within the building:

- High-piled stock
- Unpalletized storage
- Stock from shelves and the shelves
- Wet bags
- Unstable or reactive liquids or solids that have been subjected to fire
- Etiologic or radioactive material containers
- Open or closed tanks of liquids
- Safes or files on upper floors

- Heavy machinery on upper floors
- Overhead fixtures such as lights, signs or duct work
- Suspended ceilings, piping, manufacturing or processing equipment

Placing Contents In Proper Order

Restoring furnishings builds good public relations

Particularly in the residential districts, where furnishings have been arranged for covering, the contents should be replaced in as nearly the same order as they were before the fire. This should be done after the water and debris have been removed. The owner or occupant should be summoned to supervise replacing the furnishings of the rooms. This practice will serve two purposes. Not only will a better job be done, but the operation lets the owner see how the department has functioned to save his property. Extreme caution should be exercised in rearranging furnishings. Boots may need to be rinsed or removed, and sooty gloves discarded while putting household items in order. Such materials as curtains, draperies, pictures and mirrors should be hung first, for the walls are more accessible before the furniture is replaced. Remember that the emergency has passed and that carelessness at this time would, no doubt, nullify any and all desirable public relations effects previously established. As much care and precaution should be given to the furnishings of people of lesser means as should be given to the more expensive furnishings of the wealthy. Restore the heat to the building to help in drying out the contents and the building itself.

Making The Area Safe

Structural hazards should be corrected or guarded

The fire department must be concerned with removing dangerous structural conditions for the safety of their personnel and to protect the public in general. It may be necessary to shore up ceilings or interior walls, protect any holes burned in floors or stairways, check the integrity of stairs or ladders before use, and guard any sumps, open tanks or mantraps. Signs, overhangs or marquees that have damaged or weakened should be ordered removed. Consideration must be given to roof mounted equipment such as air conditioning and heating units, cooling towers, or water tanks that are causing stress or possible collapse.

The responsibility to actually remove larger hazards like dangerous walls or items which are creating an imminent danger to citizens on sidewalks or roadways, or to personnel performing final overhaul and mop-up operations belongs to the fire department. The department should seek assistance from other governmental departments or agencies to provide the heavy equipment required. Interdepartmental coordination will be necessary since cranes, bulldozers or other equipment may be required. The fire department will generally be concerned only with the safety aspects and will have the debris removed by the owner. Once the property has been turned back to the owner the fire department no longer is responsible for removing dangerous conditions. The fire department should point out such conditions, recommend, or in some cases for the public welfare, order their removal and assist in any way in directing the safe removal by the owner.

The fire department must also see to it that proper protection is provided to keep the public out of unsafe areas. During immediate operations it may be necessary to rope off an area. The park or street department should be solicited to provide barricades, traffic cones, signs and warning flashers. In some cases a guard may be required to prevent entry into unsafe areas. Again, interagency cooperation should be sought to provide guards from the police department. This is especially suitable where an arson investigation may be involved and the arson squad is a combined unit or a police function. A designated guard has the authority to prevent entrance to an area for a reasonable length of time or until the premises have been released back to the property owner or occupant. In cases where the time of guarding must extend over several days, it is advisable to check with the district attorney before allowing the owner or occupant the use of his property.

Police and fire should work together to limit access

Restoring Sprinkler Systems

Firefighters should have a *working* knowledge of sprinkler equipment and understand the functions of this important fire protection device. They must be thoroughly familiar with the particular systems in their districts in order to utilize them effectively during a fire or to assist private interests when leakage or breakdown occurs. Sprinklers can be an aid to fire departments, and they should be used as a tool for extinguishment; however, buildings may be damaged or deluged with water if the control valve is handled improperly. Sprinkler valves should be shut off to prevent excessive water damage, but an officer should never allow sprinklers to be shut off until he has ascertained that the fire is under control. A firefighter or responsible plant personnel should be assigned to stand by at the closed valve.

Firefighters should understand sprinkler system operations

Sprinkler equipment should always be placed back in service immediately after a fire has occurred. This action will afford protection against the possibility of another fire occurring and also against the added hazard of a rekindle. A fire officer should contact persons responsible for the protection of the building and have them reset the system immediately. This is especially pertinent during winter months. Personnel should be detailed to see that the sprinkler system is properly restored. Overhaul operations should be carried out as usual to see that extinguishment is complete. Since sprinkler systems are an aid to fire fighting, building owners who incur the expense of installation should receive the complete cooperation of the fire department whenever necessary.

The most common single cause of sprinkler failure is closed valves. In many of these cases, the valves are shut off prematurely by watchmen and other employees. Such persons should be trained in the proper handling of sprinkler equipment. This training rightfully comes under the classification of salvage, and although it means additional time and effort for the building owners, it will be time and effort well spent to reduce possible future losses. Fire department personnel should be taught how to efficiently support automatic sprinkler systems. A set of instructor guide sheets and color slides are available from the International Fire Service Training Association.

Restoring Standpipe Systems

Standpipe systems consist of large water pipe risers which usually extend from the lower floor of a building to either the top floor or the roof. There are hose connection outlets at each floor level, and these stations are usually strategically located for easy access to either occupants or the fire department depending on the design and intended use of the system.

Standpipe systems may be either wet or dry depending on the usage intended. Wet-pipe systems are generally considered for use by the occupants. Wet pipes contain water under pressure supplied by piping from an adequate water source. Restoring wet systems is the responsibility of management although fire department supervision and recommendations are welcomed.

Although dry-pipe systems are similar in appearance, they are designed to be supplied by fire department pumpers through siamese connections. When fire department personnel use dry-pipe standpipes to deliver water under pressure to upper floors, it is their responsibility to drain the system, close all valves and restore all outlet caps.

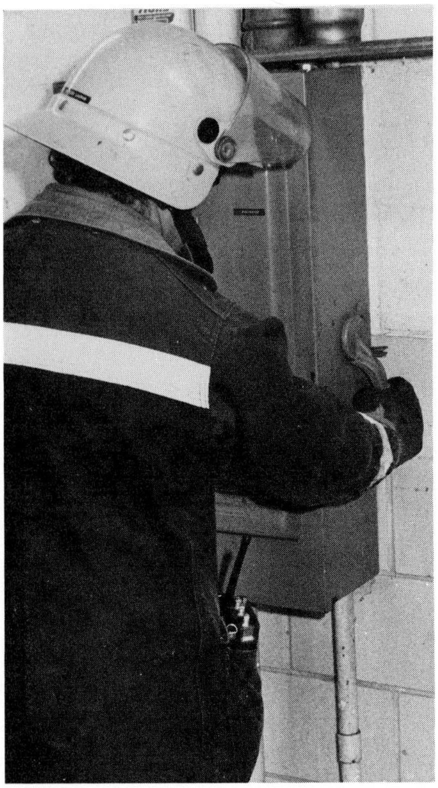

Figure 8-5 The fire department should control electrical power when it can be reached and shut down safely.

Controlling Electric Service

By far the most common utility problem is the control of the electrical system (Figure 8-5). When fire involves the structural elements of a building, the power usually must be disconnected. A good policy is to stop the flow of electricity into a building whenever the possibility exists that electrical equipment may be damaged or if electrical wiring is exposed. However, consideration must be given to the lighting needs of escaping occupants or the possibility of occupants being stranded in elevators. After an inspection of the equipment, service in most cases can at least then be reactivated to supply power for lights, tools and other operations.

Arrangements should be made in advance with the company supplying power for the immediate response of an emergency service crew to all structure fires (Figure 8-6). The utility company should understand that its personnel must be under the general supervision of the commanding fire officer, although technical supervision should come from the service crew leader. Even with the most complete planning for utility crew assistance, at times this help cannot be immediately counted upon. Stormy weather, floods, earthquakes, or other emergencies place unusually heavy workloads on these crews; their response could be delayed for some time.

Figure 8-6 If the master switch or breaker box cannot be reached to control electrical power, it is recommended that the responding utility company crew disconnect the power supply.

Controlling Gas Service

In controlling natural or liquefied petroleum gases, fire department members should be aware of the location of shutoff valves through pre-fire planning so that no delay will result (Figure 8-7). Natural gas controls are usually located at the meter of an individual occupancy. Master control valves are usually installed in multiple occupancies, but individual occupancy

Figure 8-7 The gas service can be controlled by closing the quarter turn valve on the intake line. Preplans should indicate the location of the valve.

shutoffs should be used because utility emergency crews will then spend less time and effort in restoring service. Sometimes street shutoff valves are provided on gas mains. If these have to be used for control, gas company emergency crews should handle the work under the general supervision of the commanding fire officer.

The same basic control factors apply to both natural and liquefied petroleum gases except for an important difference in the vapor density of the two gases. Under normal conditions natural gas is lighter than air and will rise and diffuse in the atmosphere. Liquefied petroleum gas is heavier than air, will hang at low levels, and will collect in basements, the bottom of vertical shafts or other depressions. Therefore, additional precautions will be necessary, including the ventilation of these areas and the elimination of potential ignition sources.

Figure 8-8 Since a leak of a drop of water per second can amount to two gallons per day and a broken pipe is equivalent to the discharge of a sprinkler head or much greater depending on the pressure and size of the pipe, any failure in the water service must be controlled immediately.

Controlling Water Service

Water damage, a major concern of fire departments, is usually associated with hoselines, sprinkler systems and other fire control devices. Additional damage is sometimes the result of a failure of domestic water systems that may occur during a fire or from other causes. Control of water services therefore may be necessary to reduce water damage (Figure 8-8).

Although a number of ways may be found to stop the flow of water within or to a structure, generally the best practice is to close valves as near the leak as possible. Most plumbing fixtures — sinks, lavatories and toilets — are supplied with water through wall-mounted valves installed under the fixtures. Water systems in large structures are usually equipped with valves that can be used to control leaks in affected sections or floors. Since sectional control valves are seldom labeled and often difficult to locate, maintenance personnel should be asked about them. Most structures, large and small, are equipped with master control valves at the point where the domestic water supply enters the structure. These valves may be installed above ground or in a box or pit near the structure.

Water meters are usually found in pits or boxes near curb lines or in basements. Most meter installations incorporate a valve on the street side of the meter. Valves are installed at intervals in street mains; utility crews can isolate sections of street main systems to control water leaks that cannot be controlled by other methods.

Deodorizing Problems

Although deodorizing is not the responsibility of the fire department, many fire departments provide this assistance as part of their public relations. Smoke odors may be reduced after some residential or commercial fires with only a reasonable amount of service time required of the firefighters. If a fire department is faced with a more complex problem of deodorizing a large building, the process may require more money, time and equipment than is usually available, and the occupant should consult an organization which specializes in this type work. Removing smoke odors can best be accomplished by the proper use of satisfactory smoke odor counteracting chemicals. Fire insurance adjusters find it difficult to evaluate smoke odor damage because smoke odors increase in proportion to the time they are permitted to remain in contact with merchandise. Likewise, the longer these odors are permitted to remain, the more difficult they are to remove. It is, therefore, vitally important to remove smoke odor from a contaminated area as soon as possible.

Companies that market odor counteractants are listed by Underwriter's Laboratories. These products in respect to their flammable characteristics are rated less hazardous than paraffin oil. They are also considered nontoxic by the Federal Food and Drug Administration. They are non-staining, when used according to instructions, colorless and have a lasting odor neutralizing ability. Most odor counteractants may be purchased in various quantities from a distributor or dealer.

Releasing The Premises

After overhaul activities are completed, releasing the premises back to the property owner is usually in order. This is a simple yet formal procedure and should be done in the presence of a witness. This release may only involve the property owner, but in some instances, other persons are involved. For this reason, public relations play an important part in making the proper release. Tactful, but firm recommendations should be made to the correct persons, and the utmost cooperation should be practiced. The following is a list of persons who may be involved and the fire department's responsibilities to them.

Work cooperatively with all parties involved

- The owner and occupants

  Advise about conditions, especially any remaining unsafe conditions.

  Recommend necessary precautions.

- Utilities

  Give full cooperation.

  Recommend necessary precautions.

- Other authorities

  Cooperate with law enforcement representatives.

  Cooperate with city officials.

  Notify proper authorities of suspicious fires.

HELPS IMPROVE PUBLIC RELATIONS

The performance of effective salvage and overhaul will show citizens that there is concern for the protection of their property. During a time of crisis the seemingly little extras are important to people and it will be returned later in good will and support. Small household momentos adequately covered and saved for a youngster or senior citizen will assist in spreading a reputation of service for the fire department. Commercial or manufacturing locations large and small can profit considerably from good salvage operations that control water damage or save particularly valuable stock, records or working documents. An aggressive salvage effort can lend support to necessary fire department budget increases. What is of value to others is hard to determine; therefore, following the basic fire service objective of saving property which implies using all salvage techniques will enhance the fire department relationship with the public.

Appendix A

The following equipment is recommended by the NFPA Committee on Automotive Fire Apparatus for salvage trucks in NFPA Standard 1901. This text should be reviewed with the equipment recommended in the various salvage and overhaul kits in IFSTA 104 to guide a fire department in its selection of salvage equipment. The equipment probably would be dispersed among engine, ladder or squad companies to meet local salvage needs.

Basic equipment on salvage trucks:

Sprinkler equipment consisting of at least 12 standard heads of assorted temperatures and types; 6 stoppers, tongs, or wedges; wrenches for the types of heads carried and an assortment of piping plugs and caps.

- 3 stillson wrenches, 10-, 14- and 16-inch sizes
- 2 monkey wrenches
- 20 salvage covers
- 2 floor runners, 3 by 18 feet
- 1 roll of 15-pound tarpaper or plastic sheeting, at least 8 mil
- 4 brooms
- 4 mops
- 2 mop wringers with buckets
- 1 water vacuum
- 6 18-inch squeegees with handles
- 3 scoop shovels
- 1 submersible pump
- 2 10-foot metal ladders with folding roof hooks
- 2 8-foot pike poles
- 125 feet of ½-inch diameter rope
- 125 feet of ⅜-inch diameter rope
- 2 6-pound axes
- 1 36-inch crowbar
- 2 pinch bars
- 2 sledge hammers
- 3 claw hammers
- 1 box of assorted nails
- 2 padlocks and hasps with keys
- 1 heavy-duty stapler
- 4 assorted hand saws
- 1 power saw
- 1 smoke ejector
- 1 2500 watt electric generator
- 3 portable 500 watt floodlights
- 6 50-foot lengths of electrical cable with locking type connectors 14-3 gage or heavier
- 3 pigtail ground adapters, 2 wire to 3 wire 14-3 gage or heavier with 12-inch minimum length
- 2 100-foot lengths of cable with locking type connectors
- 2 electric hand lights
- 1 first aid kit
- 2 approved portable fire extinguishers to handle A, B, and C class fires
- 1 roll heavy-duty cellophane tape, 2-inch width
- 1 power operated deodorizer unit
- 2 approved positive-pressure self-contained breathing apparatus with extra tanks

Appendix B

Plastic Sheeting for Salvage, Easily Applied, Has Many Uses

by Harry A. Lefebre, Battalion Chief
Warwick, Rhode Island, Fire Department

Statistics show that too many injuries occur during overhauling after a fire and a part of this evolution is the covering of openings with tar paper held in place by laths and nails. A man climbing a ladder with tar paper under one arm and carrying laths, nails and a hammer seems to be an accident looking for a place to happen.

Recognizing the accident potential of such situations, our department set out to find a better way. This greater safety impetus, coupled with new home construction methods and materials, produced a modern approach to salvage operations.

Four and 6-mil plastic sheeting available in convenient rolls, forms the nucleus of the new approach. The management of the Bostitch Division of Textron, made aware of our attempt to find the better way, donated Bostitch T-5 Tacker staple guns to our department. This was arranged by Douglas Barber, who doubles as safety engineer for Bostitch and chief of the Bradford Volunteer Fire Company. Boxes constructed by Warwick fire fighters hold all the necessary tools and equipment to perform covering operations safely and efficiently.

Materials used

The kit for covering windows and ventilation holes in roofs contains 4-mil plastic, 4 feet wide, in rolls 100 feet long. The tool inventory includes the T-5 Tacker, a knife with disposable blades, extra staples and an apron to carry the tools. The plastic is available 4 feet wide, folded in half, which reduces the size of the carrying case so that it can be carried on engines as well as ladder trucks.

Ventilation holes in roofs can be easily covered by setting the kit above the hole, pulling the plastic sheeting down over the hole and securing it with staples every 3 to 5 inches. A more secure patch can be made by folding the edges over before stapling. Experience has shown that this covering can withstand a moderate snow cover for weeks.

Covering windows is just as easy as roofs and markedly safer than the old tar paper and lath method. First-floor windows can be covered by placing the kit on the ground and pulling the plastic up over the window. Second floor and higher are covered easier by cutting a length of plastic sheeting from the roll on the ground and taking it up a ladder. We have found that by stapling every 3 to 6 inches, a secure covering is made and the building can still breathe. Plastic applied as described has withstood the fabled New England weather for weeks while thoroughly protecting interiors.

Extra lights add safety

We have found a number of advantages to this method in addition to the safety of the man assigned to covering duties. The plastic allows light to pass through and therefore makes interior mopup safer. Also, the drenched interiors dry out much more quickly than when covered with tar paper. The homeowner is much happier to have only the tiny holes made by staples in his window trim instead of unsightly marks left by nails used to hold laths in place.

In keeping with the tradition of leaving the premises in the best possible condition after a fire, plastic has proved to be more suitable than tar paper.

Our success with plastic for covering openings started us thinking of plastic for salvage covers. In addition to the 4-foot width, sheeting is also available 16 feet wide, folded in quarters to make a roll 4 feet wide. We found that one man could spread covering material, working from the kit, over even the most difficult articles of equipment. The light weight of the sheeting is particularly advantageous when covering delicate articles, and the fact that the man can see what he is covering makes him more careful. Often, one stretch of plastic can cover the contents of an entire room where formerly two or more canvas covers had to be spliced together. Being able to select the length needed to do a particular jos is a definite asset.

Better for shelves

In covering shelves full of stock, plastic once again is superior. It can be stretched quickly and gently up over the shelves and secured at the top with staples along the molding at the ceiling or along the top shelf. It is therefore unnecessary to remove the stock or to nail or hang heavy canvas covers with S hooks that too often damage stock.

The flexibility of operations using plastic seems to be endless. If a chute is needed to divert water coming through ceilings, the plastic can be stapled along the molding, stretched to a window and stapled to form a durable chute. If water is to be removed from a building down a stairway, plastic sheeting can be rolled down the stairs, in the past, this operation took a minimum of two canvas covers, spliced together and a problem usually presented itself at a turn in a stairway.

A trough to move water across an area can easily be constructed by placing wood on each side and laying the plastic over it. The sheeting can be held in place by the weight of the water or it can be stapled.

The possibilities of plastic are endless, limited only by the imagination of the men trying to solve salvage problems.

Property owners pleased

The public relations possibilities are another plus for this material. Before our shift to plastic, we were anxious to retrieve our covers a few days after a fire so that they could be cleaned, repaired and placed back in service. Unfortunately, the property owner usually had not been able to make the necessary repairs and wished to retain the covers. Often, our attempt to please the property owner resulted in severely damaged or lost covers.

With plastic, all this has changed. The covers are spread and left to be removed by property owners at their leisure. The owners' gratitude sometimes takes the form of offering to pay for rolls of replacement plastic.

Frequently, there are no adequate facilities to clean, dry and repair covers. Often covers are washed on the station ramp and hung over reserve apparatus or suspended from the station ceiling to dry. Trying to protect the "salvage side" of covers is virtually impossible and the next time they are used, damage is done to articles you are trying to protect. Good salvage cover maintenance also requires periodic unfolding and airing. Plastic sheeting involves none of these drawbacks.

Withstands most chemicals

The fact that plastic is impervious to most chemicals is another advantage in this day and age when chemicals once found only in industry are readily found in homes. Plastic covering can be damaged by sparks and embers, but the economic loss involved is quite small when compared to that of canvas salvage covers.

From the standpoint of economics, plastic is cheaper than tar paper and laths for covering operations. The cost of a roll of plastic wide enough for salvage covers is one-third that of one salvage cover and provides from five to six "covers" of comparable size.

Since our shift to plastic, we have had many inquiries about our kits. We have prepared drawings of the salvage kits and explanatory material that are available upon request.

Appendix C
GLOSSARY

ACCORDION FOLD
A method of folding a salvage cover, which when completed resembles the bellows of an accordion.

AUGER
A tool for boring holes in floors.

ARSON
The willful burning of a dwelling, building, structure or other property, including one's own.

BALE HOOK
Useful tool in moving bales or boxed goods around, and overhauling stuffed furniture or mattresses.

BALLOON THROW
A method of spreading a salvage cover over material that utilizes air trapped under the cover to float it in place.

CARRYALL
Waterproof carrier or bag used to carry and catch debris or used as a water sump for immersing small burning objects.

CATCHALLS
A retaining basin usually made from salvage covers to impound water dripping from above.

CATCH BASIN
A device made from covers to trap considerable amounts of water by forming a reservoir from chairs, tables or any suitable material available.

CHUTE
A salvage cover arrangement that channels excess water from a building.

CONCEALED SPACE
Space hidden between floors, ceilings, roofs and walls in which fire can spread easily and undetected.

CONTENTS
Furnishings, merchandise and any machinery or equipment not part of the building structure.

CROSSOVER THROW
A method of throwing a cover in which one edge of the salvage cover is crossed over the other edge prior to the throw. This arrangement allows both the operators to throw with their right hand and to use the bottom edge as a booster for the throwing edge. The crossover throw is especially adapted to covering high objects.

DOUBLE-EDGE SNAP THROW
A method of throwing a cover, similar to the single edge snap throw; however, it is designed to cover two relatively narrow objects located close to each other.

FLOOR RUNNERS
Heavy plastic or canvas material usually three feet by 18 feet placed on floors or carpeting to prevent damage from mud, water and debris damage.

GROMMETS

The reinforced eyelets in salvage covers through which fasteners are passed that permit hanging of the covers over shelving, on walls, with ropes or pike poles.

IKE-O-HOOK

An iron hook with an eyelet on the other end used for hanging covers from pike poles or ropes. They can also be used to keep covers taut that are being used as chutes and to suspend electrical cords above wet floors.

INFRARED VIEWER

A device that detects infrared energy that is emitted by heated objects. The viewer is primarily used by the fire service to detect hidden fires.

LAKE TEST

A method of testing a cover for leaks by utilizing it as a catchall to form a "lake" of water and holding it for 12 hours without leaking.

LATH

A thin, narrow strip of wood used to hold covers, sheeting or tarpaper in place.

MATTRESS CHAINS

Light chains approximately eight feet long with hooks or locking devices. They are used to keep a mattress rolled while removing it from the premises.

MOPS AND WRINGER

Tools used for cleanup and removal of water.

OVERHAUL

Those operations which consist of searching for and extinguishing hidden or remaining fires; placing the building and its contents and the fire area in a safe condition; determining the cause of the fire; and recognizing and preserving evidence of arson.

PIPE PLUGS AND CAPS

An assortment of plugs and caps that can be used to stopper broken water lines to minimize water damage.

PORTABLE PUMP

A small gasoline or electric driven pump that can be easily moved throughout the building by one or two persons.

PUNCH AND CHISEL

Tools that can be used to make holes for drainage.

REKINDLE

The re-ignition of material after extinguishment.

"S" HOOKS

Cold rolled steel formed into the shape of a "S" with the ends sharpened to a point. "S" hooks can be driven into a wall and then used to hang salvage covers.

SALVAGE

Those methods and operating procedures which further reduce fire, water and smoke damage during and after fires.

SALVAGE COVER

Covers that are primarily designed to protect building contents from water damage during extinguishment of a fire. Usually 14 feet by 18 feet or 12 feet by 18 feet in size.

SASH CORD
Used for tie backs, guide lines or securing covers by lacing it through the grommets.

SCUPPER
A form of drain opening provided near the floor level to remove water through the walls to the exterior of a building in order to reduce possible water damage from hose streams and sprinklers.

SEWER DRAIN GUARD
A strainer to prevent debris from getting into the sewer system when using soil pipes to remove water during salvage operations.

SINGLE-EDGE SNAP THROW
A method of throwing a cover by a snap action, designed to be used in narrow spaces.

SIPHON EDUCTOR
An eductor that is helpful in removing water from basements, sumps or low areas.

SMOKE CURTAINS
Salvage covers placed in stairways, halls or doors to assist in the prevention of smoke spread.

SPLICING
The process of combining two covers into a larger or longer one with a leak-proof seal.

SPONGES AND CHAMOIS
Can be used for drying furniture and contents, or removing small amounts of water.

SPRINKLER HEAD
A device that basically consists of a fusible link, deflector and a threaded pipe nipple. The head is screwed into a pipe that supplies water exclusively for fire protection. In the presence of fire the fusible link melts and releases the water which is broken into fine water droplets by the deflector to control the fire.

SPRINKLER RISER
A vertical pipe used to carry water to the sprinkler system.

SPRINKLER STOPPER (TONGS)
A device used to stop the flow of water from individual sprinkler heads.

SPRINKLER WEDGE
A piece of soft wood in the shape of a wedge, to be used to stop water from flowing out of a sprinkler head.

SPRINKLER WRENCH
A special wrench that is designed to be used for sprinkler heads.

SQUEEGEE
A rubber-edged broom-like device used in salvage to assist in the removal of water from floors by pushing the water to a drain, disposal area or place where it can be picked up.

STANDPIPE SYSTEM
A network of pipes used to supply water to fire hoses in buildings.

SUMP
A pit or reservoir serving as a receptacle for water often connected to a drain.

WATER VACUUM
Electrical-powered vacuum device used in salvage to pick up remaining water.

Appendix D

The Metric System Of Measurement

Since the development of the metric system in the eighteenth century, the vast majority of the world's countries have gone metric. In fact, only six countries, in 1974, were uncommitted to the metric system.

There are two basic reasons for the conversion of countries to the metric system. The first reason for going metric is for simplicity and efficiency. The metric system is based upon simple powers of ten where mass, volume and length are all interrelated in the system of measures. The fact that the power of ten is used in metrics makes computations simple, with a reduction in errors and a savings of time.

The second fundamental reason for adopting the metric system of measurement is fairly obvious considering the fact that almost all countries were currently using metrification. The simple economics of world trade displays the advantages and necessity of using a common system of measurement. Since the majority of countries are using metrics, they supply products in metric measurements. These same countries, as consumers, will often demand products in metrics. In order to remain economically competitive in the world market the need for metrification is evident.

Since its inception, the metric system of measurement has not been a static one. The system has continued to be refined and, as a result, the ultimate in metrification has been developed. The International System of Units, simply referred to as SI, was devised in 1960 by the metric nations. Already adopted by more than 30 countries, this system embodies the latest in metric study.

UNITS OF METRIC SYSTEM

Length

The basic unit of the metric system is the meter. In establishing the length of the meter, the developers wanted to use a basic fact of nature. After several changes the meter was finally defined as "1,650,763.73 wave lengths of the orange-red line of krypton 86". This definition is based on a physical condition which will always be the same and is a true standard of length. From the basic unit of the meter, the system of lengths is simply expanded in multiples of tens in either direction. By using powers of ten, confusing fractions are eliminated since all lengths can be simply expressed using decimals. The length measurements in metrics and English equivalents are:

MULTIPLIER		UNIT OF LENGTH	SYMBOL		OUR EQUIVALENT
1/1000 meter	=	1 millimeter	mm	=	0.0393701 inch
10 millimeters	=	1 centimeter	cm	=	.3937 inch
10 centimeters	=	1 decimeter	dm	=	3.937 inches
10 decimeters	=	1 meter	m	=	1.09361 yds.
10 meters	=	1 decameter	dam	=	32.81 ft.; 10.936 yds.
10 decameters	=	1 hectometer	hm	=	328.00 ft.; 109.3 yds.
10 hectometers	=	1 kilometer	km	=	0.621371 miles

LENGTH

TO CONVERT	TO	MULTIPLY BY
inches	centimeters (cm)	2.54
feet	meters (m)	0.3048
yards	meters (m)	0.9144
miles	kilometers (km)	1.852

As you will notice, the word meter is used in developing all other words. The prefixes used are either Greek or Latin words or roots. The same prefixes are used in developing other measurements — again, a simplification. Since metrics is in powers of ten, it is extremely simple to make other conversions. For instance, how many millimeters are equal to an inch? Knowing that 2.54 cm = 1 inch and knowing that 10 mm = 1 cm, it is simple to deduce 1 inch = 25.4 mm.

Metric Squares

The method for devising square-measures is simple. For example, begin with a square land plot measuring one meter per side. This is a square meter (m^2). A square this size is also equal to one which measures 10 dm by 10 dm (since 10 dm = 1 m) and therefore is equal to 10 dm × 10 dm = 100 square decimeters (dm^2). We can then say that 1 m^2 = 100 dm^2.

MULTIPLIER		UNIT OF AREA	SYMBOL		OUR EQUIVALENT
—	=	1 sq. millimeter	mm^2	=	0.00155 sq. in.
100 mm^2	=	1 sq. centimeter	cm^2	=	.155 sq. in.
100 cm^2	=	1 sq. decimeter	dm^2	=	15.5 sq. in.
100 dm^2	=	1 sq. meter	m^2	=	1.196 sq. yd.
100 m^2	=	1 sq. decameter	dam^2	=	0.025 acre
100 dam^2	=	1 sq. hectometer	hm^2	=	2.471 acres
100 hm^2	=	1 sq. kilometer	km^2	=	.386 sq. mi.

One hm^2 is often referred to as a hectare (ha).

AREA

TO CONVERT	TO	MULTIPLY BY
square inches	square centimeters (cm^2)	6.452
square feet	square centimeters (cm^2)	929.030
square yards	square meters (m^2)	0.836
acres	hectares	0.405
square miles	hectares	258.999
square miles	square kilometers (km^2)	2.59

Metric Cubes

To measure volume or capacity, develop a cube 1 meter by 1 meter by 1 meter. Its volume is a cubic meter (m^3). This cube is the same size as a cube measuring 10 dm to a side, and therefore has the volume of $10 \text{ dm} \times 10 \text{ dm} \times 10 \text{ dm} = 1000 \text{ dm}^3$. Consequently, there are 1000 dm^3 in 1 m^3. A cubic metric-unit is always 1000 times larger than the next smallest one.

MULTIPLIER		UNIT OF VOLUME	SYMBOL		OUR EQUIVALENT
—	=	1 cubic millimeter	mm^3	=	6.1×10^{-5} cubic inch
1000 mm^3	=	1 cubic centimeter	cm^3	=	.061 cubic inch
1000 cm^3	=	1 cubic decimeter	dm^3	=	61.023 cubic inches
1000 dm^3	=	1 cubic meter	m^3	=	1.30795 cubic yards
1000 m^3	=	1 cubic decameter	dam^3	=	
1000 dam^3	=	1 cubic hectometer	hm^3	=	
1000 hm^3	=	1 cubic kilometer	km^3	=	

VOLUME: CUBIC

TO CONVERT	TO	MULTIPLY BY
cubic inches	cubic centimeters (cm^3)	16.387
cubic feet	cubic decimeters (dm^3)	28.317
cubic yards	cubic meters (m^3)	0.7645

VOLUME AND CAPACITY

The basic unit of measure for volume or capacity is the liter. The liter is defined as 1 kilogram of pure water at 4°C. The liter is also equal in volume to one cubic decimeter (1 dm^3); hence the interrelationship of measurement of volume and length. To expand the metric volume, all that is needed is to expand the liter by multiples of 10 in either direction and apply the normal prefixes.

MULTIPLIER		UNIT OF CAPACITY	SYMBOL		EQUIVALENT LINEAR MEASURES
1/1000 liter	=	1 milliliter	ml	=	1 cubic centimeter
10 milliliters	=	1 centiliter	cl	=	10 cubic centimeters
10 centiliters	=	1 deciliter	dl	=	100 cubic centimeters
10 deciliters	=	1 liter	l	=	1 cubic decimeter
10 liters	=	1 decaliter	dal	=	10 cubic decimeters
10 decaliters	=	1 hectoliter	hl	=	100 cubic decimeters
10 hectoliters	=	1 kiloliter	kl	=	1 cubic meter

The English conversions are as follows:

TO CONVERT	TO	MULTIPLY BY
ounces	milliliters (ml)	296.00
pints	liters (l)	.4732
quarts	liters (l)	.9464
gallons	liters (l)	3.785

The most commonly used conversion will be in converting gallons per minute to liters per minute: 1 gpm = 3.785 liters/minute (lpm).

METRIC MASS

The basic metric unit for mass measurement is the gram. It has already been discussed that 1 cubic centimeter = 1 milliliter and furthermore, 1 milliliter has a volume equal to that of 1 gram of water. Again the measurements are interrelated. By multiples of ten, the metrics for mass include:

MULTIPLIER		UNIT OF MASS	SYMBOL		OUR EQUIVALENT
1/1000 gram	=	1 milligram	mg	=	0.015 grain
10 milligrams	=	1 centigram	cg	=	.154 grains
10 centigrams	=	1 decigram	dg	=	1.543 grains
10 decigrams	=	1 gram	g	=	.035 ounces
10 grams	=	1 decagram	dag	=	.353 ounces
10 decagrams	=	1 hectogram	hg	=	3.527 ounces
10 hectograms	=	1 kilogram	kg	=	2.205 pounds

One kilogram is further defined as the mass of 1 liter of pure water at 4°Celsius.

1 kg = 1000 g
1 l = 1000 ml

but 1 liter of water has a mass of 1 kg and 1000 ml = 1 dm^3, therefore, 1 dm^3 of water has a mass of 1000 g.

TO CONVERT	TO	MULTIPLY BY
ounces	grams (g)	28.35
pounds	grams (g)	453.592
pounds	kilograms (kg)	0.454

MEASURING TEMPERATURE

The Celsius, or Centrigrade, system of heat measurement has also been adopted by most countries using metrics. In this system, the freezing point of water is 0° and the boiling point of water is 100°. The difference is 100° which fits in well with the decimal notation of the metric system. To convert Celsius to Fahrenheit, using °C = 5/9(°F − 32). To convert the other way, using °F = 9/5(°C + 32).

PASCAL

One other metric unit that is particularly important to firefighters is the unit to measure pressure, the pascal (Pa).

1 psi = 6894.757 Pa, so 1 psi = 6.895 kPa

Metric Equivalents

PAGE	ENGLISH		METRIC
	CHAPTER 2		
13	9 by 12 feet	=	2.77 m by 3.66 m
13	14 by 18 feet	=	4.27 m by 5.49 m
14	2 by 6 inches	=	5.08 cm by 15.24 cm
14	5 inches	=	12.7 cm
17	2 inches	=	5.08 cm
17	3 inches	=	7.62 cm
22	3 feet	=	.914 m
22	20 to 30 feet	=	6.1 m to 9.14 m
22	8 inches	=	20.32 cm
22	10 inches	=	25.4 cm
23	1/4 or 5/16 inch	=	.64 cm or .79 cm
23	8 inches	=	20.32 cm
	CHAPTER 3		
29	12 inches	=	30.48 cm
34	4 inches	=	10.16 cm
38	10 inches	=	25.4 cm
38	3 feet	=	.914 cm
40	9 inches	=	22.5 cm
40	18 inches	=	45.7 cm
41	3/8 by 9 by 35 inches	=	.95 cm by 22.8 cm by 88.9 cm
57	3 or 4 feet	=	.914 m or 1.22 m

PAGE	ENGLISH		METRIC
	CHAPTER 4		
66	20 or 24 feet	=	6.10 m or 7.32 m
69	100 gallons	=	379 l
70	3 feet	=	1 m
71	1/8 inch	=	.32 cm
72	2 feet	=	.61 m
72	1 1/2 feet	=	.46 m
74	10 feet	=	3.05 m
78	12 inches	=	30.48 cm
79	20 inches	=	50.8 cm
80	10 inches	=	25.4 cm
81	18 inches	=	45.7 cm
81	8 inches	=	20.32 cm
83	6 to 8 feet	=	1.83 m to 2.44 m
83	14 to 18 feet	=	4.27 m to 5.49 m
	CHAPTER 5		
87	250 gpm	=	947.5 lpm
87	ton (2000 pounds)	=	.9 metric ton (907 kilograms)
91	350 gpm	=	1326.5 lpm
95	1 1/2 inches	=	3.81 cm
95	2 1/2 inches	=	6.35 cm
	CHAPTER 6		
106	6 or 8 feet	=	1.82 m or 2.44 m
107	6 or 8 feet	=	1.82 m or 2.44 m
109	8 feet	=	2.44 m
	CHAPTER 7		
115	2 1/2 inches	=	6.35 cm
115	1 1/2 inches	=	3.81 cm
116	1 1/2 inches	=	3.81 cm
	CHAPTER 8		
128	6 inches	=	15.24 cm
129	6 inches	=	15.24 cm
136	2 gallons	=	7.58 l

PAGE	ENGLISH		METRIC
	APPENDIX A		
139	10 inches	=	25.4 cm
139	14 inches	=	35.6 cm
139	16 inches	=	40.6 cm
139	3 by 18 feet	=	.914 m by 5.49 m
139	15 pounds	=	6.8 kg
139	18 inches	=	45.7 cm
139	10 feet	=	3.05 m
139	8 feet	=	2.44 m
139	125 feet	=	38.13 m
139	1/2 inch	=	1.27 cm
139	3/8 inch	=	.95 cm
139	6 pounds	=	2.7 kg
139	36 inches	=	91.44 cm
139	50 feet	=	15.25 m
139	12 inches	=	30.48 cm
139	100 feet	=	30.5 m
139	2 inches	=	5.08 cm
	APPENDIX B		
140	4 feet	=	1.22 m
140	100 feet	=	30.5 m
140	3 to 5 inches	=	7.62 cm to 12.7 cm
140	3 to 6 inches	=	7.62 cm to 15.24 cm
140	16 feet	=	4.88 m
	APPENDIX C		
142	3 by 18 feet	=	.914 m by 5.49 m
143	14 by 18 feet	=	4.27 m by 5.49 m
143	12 by 18 feet	=	3.66 m by 5.49 m

THE INTERNATIONAL FIRE SERVICE TRAINING ASSOCIATION

The International Fire Service Training Association is an educational alliance organized to develop training material for the fire service. The annual meeting of its membership consists of a workshop conference which has several objectives —

. . . to develop training material for publication
. . . to validate training material for publication
. . . to check proposed rough drafts for errors
. . . to add new techniques and developments
. . . to delete obsolete and outmoded methods
. . . to upgrade the fire service through training

This training association was formed in November 1934, when the Western Actuarial Bureau sponsored a conference in Kansas City, Missouri, to determine how all agencies that were interested in publishing fire service training material could coordinate their efforts. Four states were represented at this conference and it was decided that, since the representatives from Oklahoma had done some pioneering in fire training manual development, other interested states should join forces with them. This merger made it possible to develop nationally recognized training material which was broader in scope than material published by an individual state agency. This merger further made possible a reduction in publication costs, since it enabled each state to benefit from the economy of relatively large printing orders. These savings would not be possible if each individual state developed and published its own training material.

From the original four states, the adoption list has grown to forty-four American States; six Canadian Provinces; the British Territory of Bermuda; the Australian State of Queensland; the International Civil Aviation Organization Training Centre in Beirut, Lebanon; the Department of National Defence of Canada; the Department of the Army of the United States; the Department of the Navy of the United States; the United States Air Force; the United States Bureau of Indian Affairs; the United States General Services Administration, and the National Aeronautics and Space Administration (NASA). Representatives from the various adopting agencies serve as a voluntary group of individuals who govern policies, recommend procedures, and validate material before it is published. Most of the representatives are members of other international fire protection organizations and this meeting brings together men from several related and allied fields, such as:

. . . key fire department executives and drillmasters,
. . . educators from colleges and universities,
. . . representatives from governmental agencies,
. . . delegates of firefighter associations and organizations, and
. . . engineers from the fire insurance industry.

This unique feature provides a close relationship between the International Fire Service Training Association and other fire protection agencies, which helps to correlate the efforts of all concerned.

The publications of the International Fire Service Training Association are compatible with the National Fire Protection Association's Standard 1001, "Fire Fighter Professional Qualifications (1974)," and the International Association of Fire Fighters/International Association of Fire Chiefs "National Apprenticeship and Training Standards for the Fire Fighter." The standards are an effort to attain professional status through progressive training. The NFPA and IAFF/IAFC Standards were prepared in cooperation with the Joint Council of National Fire Service Organizations of which IFSTA is a member.

In order to formally adopt IFSTA training material, the firefighters should first request it through their training agencies. While adopting does not obligate any sponsoring agency, it still profits by sending a representative to the Annual IFSTA Validation Conference. This representation permits a voice and a vote on the ever-changing policies, techniques, and procedures which must be reflected in each publication. Adoption further adds prestige to any training program and this, in turn, adds prestige to IFSTA publications.

The International Fire Service Training Association meets each July at Oklahoma State University, Stillwater, Oklahoma. Fire Protection Publications at Oklahoma State University publishes all IFSTA training manuals and texts. This department is responsible to the executive board of the association. While most of the IFSTA training manuals can be used for self-instruction, they are best suited to group work under a qualified instructor.

IFSTA CONFERENCE ATTENDEES AT THE TIME OF VALIDATION

FIRE ASSOCIATIONS

Tom Admire
Engineer
Montgomery County Fire Marshal's Office
Bethesda, Maryland

Steve Ashby
Firefighter
Department of Fire Services
Stillwater, Oklahoma

Paul Benton
President
Arkansas Firefighter's Association
Camden, Arkansas

Paul Boecker
Fire Chief
Lisle-Woodridge Fire District
Lisle, Illinois

Howard Boyd
Fire Marshal
Metropolitan Fire Department
Nashville, Tennessee

David Briley
Captain
Fire Department
Ames, Iowa

Louis Bunch
Executive Director
Oklahoma State Firefighter's Association
Oklahoma City, Oklahoma

Richard Calhoun
Fire Protection Inspector
State Fire Marshal's Office
Shreveport, Louisiana

George Carney
Chief Training Officer
Fire Department
Shreveport, Louisiana

Bob Caron
Fire Instructors Association of Minnesota
Detroit Lakes, Minnesota

Sherman Carthen
Drillmaster
Fire Department
Oklahoma City, Oklahoma

James Coder
Training Officer
Fire Department
Kalamazoo, Michigan

Ronny Coleman
Fire Chief
Fire Department
San Clemente, California

John Collins
Nashville, Tennessee

Dan Cotten
Training Division
Fire Department
Shreveport, Louisiana

Emmett Cox
Technical Consultant
International Association of Firefighters
Dolton, Illinois

Bill Darin
Fire Marshal
Fire Department
Edmond, Oklahoma

Lee Daugherty
Assistant Chief
Fire Department
Edmond, Oklahoma

Donald Davis
Indianapolis, Indiana

Robert Ely
Chief
San Diego, California

Tim Ford
Fire Department
Nichols Hills, Oklahoma

Douglas Forsman
Senior Fire Service Specialist
National Fire Protection Association
Boston, Massachusetts

Larry Francis
Fire Department
Oklahoma City, Oklahoma

Len Gatin
Deputy Chief
Fire Department
Kamloops, B.C., Canada

Ray Gish
Executive Manager
State Firemen's & Fire Marshal's Association
Austin, Texas

Ron Hamm
Indiana Fire Instructors Association
Fort Wayne, Indiana

Eric Haussermann
Lieutenant
Fire Department
Champaign, Illinois

John Hendrix
Chief Training Officer
Fire Department
Hutchinson, Kansas

Charles Henrici
Illinois Fire Inspectors
Elk Grove Village, Illinois

Robert Higgs
Assistant Training Officer
Fire Department
Topeka, Kansas

Curtis Holter
Fire Instructors Association of Minnesota
Minnetonka, Minnesota

Earl Hood
Aberdeen, South Dakota

Lonnie Jackson
Fire Inspector
Mount Prospect Fire Department
Arlington Heights, Illinois

Thomas Jefferson
Ft. Smith, Arkansas

E. F. Kaughman
Calgary, Alberta, Canada

C. H. Kennedy, Jr.
Training Officer
Fire Department
Baton Rouge, Louisiana

Raymond Linville
Indiana Fire Instructors Association
Lafayette, Indiana

Cathy Lohr
Fire Education Specialist
Guilford County Fire Marshal's Office
Raleigh, North Carolina

Edward H. McCormack, Jr.
Executive Secretary
International Society of Fire Service Instructors
Hopkinton, Massachusetts

Carl McCoy
Division Director
State Fire Marshal's Office
Springfield, Illinois

Max McRae
District Chief
Houston Fire Department
Bellaire, Texas

Richard Mazzone
Engineer
Fire Department
Victorville, California

Ron Miller
Fire Chief
Department of Fire Service
Stillwater, Oklahoma

Kenneth Mitten
Battalion Chief
Fire Department
Merced, California

Dennis Mosser
West Columbia, South Carolina

Bobby Mowles
Fire Department
Oklahoma City, Oklahoma

James Neelon
Assistant Chief
Cleveland Fire Department
Willowick, Ohio

Wayne Nelson
Sierra Volunteer Fire Department
Roswell, New Mexico

Elton Nixon
Fire Inspector
Stillwater, Oklahoma

Leonard O'Daniell
Oklahoma State Firefighters Association
Oklahoma City, Oklahoma

Dennis Ozment
Captain
Minneapolis Fire Department
Rosemount, Minnesota

James Parman
Fire Safety Inspector
Greeneville, Tennessee

Don Perry
Battalion Chief
Santa Barbara County Fire Department
Santa Barbara, California

Robert Porter
Assistant Chief
Bancroft Fire Department
Lakewood, Colorado

Roy Richardson
2nd Vice President
Indiana Fire Instructors Association
Henryville, Indiana

Charles Scott
Captain
Fire Department
Lubbock, Texas

Jim Simmons
Assistant Chief
Fire Department
Marshalltown, Iowa

Jerry Smith
Fire Department
Oklahoma City, Oklahoma

Carl Standefer
Fire Services Chief
State Fire Marshal's Office
Santa Fe, New Mexico

Donald Strahl
Indiana Fire Instructors Association
Indianapolis, Indiana

Bill Vandevort
Captain
Fire Department
Napa, California

Thomas Walsh
Training Chief
DFW Airport
Dallas/Fort Worth, Texas

Benjamin Wells
Fire Department
Hampton, Virginia

Jack Whisman
Captain
Fire Department
Longmont, Colorado

Lou Witzeman
Rural/Metro Fire Department
Scottsdale, Arizona

Earl Woolrich
Calgary, Alberta, Canada

Thomas Woods
Training Officer
Fire Department
Baton Rouge, Louisiana

EDUCATIONAL AGENCIES

Lester Ables
Field Instructor
Arkansas Fire Training Academy
East Camden, Arkansas

Louis Amabili
Director
Delaware State Fire School
Dover, Delaware

Elmo Anderson
Field Instructor
Arkansas Fire Training Academy
West Memphis, Arkansas

Stan Babin
Coordinator
Firemen Training Program
Louisiana State University
Baton Rouge, Louisiana

Charlotte Badgett
Fire & Safety Inspector
University of Texas
Austin, Texas

Jimmie Badgett
Training Consultant
Commission on Fire Protection Personnel Standards and Education
Austin, Texas

David Ballenger
Fire Service Training
Oklahoma State University
Stillwater, Oklahoma

Wesley Beitl
Technical Extension Specialist
Oklahoma State University
Stillwater, Oklahoma

F. Daniel Belanger
Coordinator II
Firemen Training Program
Louisiana State University
Baton Rouge, Louisiana

Fred W. Bell, Jr.
Fire Service Instructor
University of Kansas
Lawrence, Kansas

Ivan Blanco
Stillwater, Oklahoma

Larry Borgelt
Associate Professor
Fire Protection & Safety Engineering Technology
Oklahoma State University
Stillwater, Oklahoma

Glenn Boughton
Fire Service Instructor
The University of Kansas
Lawrence, Kansas

Gerald Brinkman
Coordinator
Fire Service Training
Howard, South Dakota

Pat D. Brock
Assistant Professor
Fire Protection & Safety Engineering Technology
Oklahoma State University
Stillwater, Oklahoma

Michael Brooks
Research Technician
Fire Protection Publications
Oklahoma State University
Stillwater, Oklahoma

E. T. Brown
Research Technician
Fire Protection Publications
Oklahoma State University
Stillwater, Oklahoma

Stephen Coan
Deputy Chief
Massachusetts Firefighting Academy
Sudbury, Massachusetts

Owen Collins
Technical Instructor
Fire Service Training
Jackson, Tennessee

Dale Cozad
Chairman
Fire Prevention & Control
Eastern Kentucky University
Richmond, Kentucky

Ray Davidson
Stillwater, Oklahoma

Joseph Donovan
Chief of Fire Training
Massachusetts Firefighting Academy
Sudbury, Massachusetts

James Doyle
Senior Instructor
Nebraska Fire Service
Lexington, Nebraska

J. F. Drake
Senior Fire Instructor
Alabama State Fire College
University, Alabama

Gene Duncan
Research Technician
Fire Service Training
Oklahoma State University
Stillwater, Oklahoma

Howard Eskridge
Fire Service Institute
University of Illinois
Urbana, Illinois

Garland Fulbright
Executive Director
Commission on Fire Protection Personnel Standards and Education
Austin, Texas

Richard Giles
University Director of Safety
Oklahoma State University
Stillwater, Oklahoma

Ronald Graw
Lieutenant
Rockford Fire Training Academy
Rockford, Illinois

Arlen Gross
Coordinator
Firemen Training Program
Louisiana State University
Baton Rouge, Louisiana

Gordon W. Harris, Jr.
Fire Service Coordinator
Waukesha County Technical Institute
Pewaukee, Wisconsin

James Hebert
Coordinator
LSU Firemen Training
Louisiana State University
Baton Rouge, Louisiana

A. E. (Gene) Hendricks
North Little Rock, Arkansas

Carrol Herring
Department Head
Firemen Training Program
Louisiana State University
Baton Rouge, Louisiana

Simon Hertzler
Vocational Education Consultant
Fire Service Training
Madison, Wisconsin

John Hoglund
Director
Maryland Fire & Rescue Institute
University of Maryland
College Park, Maryland

John Horn
Fire Service Instructor
Montana Fire Service Training School
Great Falls, Montana

Frank Hubble
Area Supervisor
Fire Service Training
Commonwealth of Virginia
Roanoke, Virginia

Everett Hudiburg
Editor Emeritus
Fire Protection Publications
Oklahoma State University
Stillwater, Oklahoma

George Hunt
Instructor
Greeneville, Tennessee

Jesse Jackson
Coordinator
Maryland Fire & Rescue Institute
College Park, Maryland

Dale Janes
Head
Fire Protection & Safety Engineering Technology
Oklahoma State University
Stillwater, Oklahoma

Thomas Johnson
Curriculum Writer
Fire Fighters Training Council
Lansing, Michigan

Tom Jones
Field Instructor
Fire Service Training
Oklahoma State University
Stillwater, Oklahoma

Jerry Laughlin
Supervisor
Fire Protection Publications
Oklahoma State University
Stillwater, Oklahoma

Joe Lewand
Manager
Colorado Fire Service Training
Golden, Colorado

Lester Lukert
Senior Fire Instructor
Nebraska Fire Service
Seward, Nebraska

Richard McIntyre
Fire Coordinator
North Carolina Department of Community Colleges
Burlington, North Carolina

Donald McKay
Fire Instructor
Minnesota Department of Education
St. Paul, Minnesota

Harold Mace
Manager
Fire Service Training and Fire Protection Publications
Oklahoma State University
Stillwater, Oklahoma

Joe Martinez
Research Technician
Fire Service Training
Oklahoma State University
Stillwater, Oklahoma

Gerald Monigold
Director
Fire Service Institute
University of Illinois
Champaign, Illinois

Jerry Moore
Mississippi State Fire Academy
Jackson, Mississippi

J. Fred Myers
Fire Service Instructor
University of Kansas
Lawrence, Kansas

Connie Osterhout
Assistant Editor
Fire Protection Publications
Oklahoma State University
Stillwater, Oklahoma

Charles Page
Associate Training Specialist
Fire Protection Training Division
Texas A & M University
College Station, Texas

Ronald Paubel
Fire Service Specialist
Fire Training
University of Missouri
Columbia, Missouri

George Pelissier
Arkansas Fire Training Academy
East Camden, Arkansas

Pedro Perez
Stillwater, Oklahoma

Tim Perkins
Research Technician
Fire Protection Publications
Oklahoma State University
Stillwater, Oklahoma

Shelton Riggs
Field Instructor
Arkansas Fire Training Academy
East Camden, Arkansas

Bob Roberts
LSU Firemen Training
Louisiana State University
Olla, Louisiana

Ralph Scantlin
Arkansas Fire Training Academy
Little Rock, Arkansas

Don Seelig
Field Instructor
Fire Service Training
Oklahoma State University
Stillwater, Oklahoma

Jerry Simmons
EMS Program Coordinator
Arkansas Fire Training Academy
East Camden, Arkansas

James Smalley
Training Officer
Arkansas Fire Training Academy
East Camden, Arkansas

Carol Smith
Administrative Supervisor
Fire Protection Publications
Oklahoma State University
Stillwater, Oklahoma

Henry Smith
Chief
Firemen's Training School
Texas A & M University
College Station, Texas

W. W. Smith
Specialist
State Department of Education
University of Alabama
Adamsville, Alabama

Edward Steiner
Assistant Professor & Head
Fire Protection Technology
OSU Technical Institute
Oklahoma City, Oklahoma

Jim Straseske
Instructor
Fire Service Institute
University of Illinois
Champaign, Illinois

Roger Sweet
Field Instructor
Fire Service Extension
Iowa State University
Ames, Iowa

Larry Thomas
Fire Training Specialist
University of Missouri
Columbia, Missouri

Max Thomas
Director
Fire Service Training
The University of Kansas
Lawrence, Kansas

Charles Tice
Program Specialist
Georgia Fire Academy
Marietta, Georgia

Glenn Wagner
Fire and Safety Officer
Texas Department of Mental Health and Mental Retardation
Vernon, Texas

Bill Warren
Staff Instructor
Mississippi State Fire Academy
Pearl, Mississippi

David Williams
Research Technician
Fire Protection Publications
Stillwater, Oklahoma

Dale Wilson
Fire Protection Specialist
Safety Department
Oklahoma State University
Stillwater, Okahoma

GOVERNMENTAL AGENCIES

Curtis Bell
Fire Protection Engineer
Tennessee Valley Authority
Chattanooga, Tennessee

Scott Bell
Firefighter
Vance AFB/Northrop
Enid, Oklahoma

Robert Bloom
Fire Control Service
Texas Forest Service
Lufkin, Texas

Robert Borges
Firefighter
McConnell AFB
Wichita, Kansas

Thomas Campbell
Fire Marshal Division
Michigan State Police
Lansing, Michigan

J. Humberto S. Guevara
Third Chief
National Institute of Fire Prevention and Protection of Nicaragua
Managua, Nicaragua

William Hanbury
Director
Interagency Coordination
National Fire Prevention and Control Administration
Washington, D. C.

William Hewitt
Assistant Fire Commissioner
Department of Labour
Brandon, Manitoba, Canada

Julian King
Fire Chief
U. S. Army Ammunition Plant
McAlester, Oklahoma

Thomas Makey
Deputy Fire Commissioner
Edmonton, Alberta, Canada

Richard Pippenger
Regional Safety Officer
Bureau of Reclamation
Boise, Idaho

Richard Rightmer
Fire Chief
Naval Weapons Center
China Lake, California

Jim Sorenson
Forest Service Cooperative Fire Protection
U. S. Department of Agriculture
Atlanta, Georgia

INSURANCE SERVICES ORGANIZATIONS

Marvin Austin
Supervisor of Public Protection
Insurance Services Office of Illinois
Chicago, Illinois

K. G. Hulme
Assistant Manager
Insurance Services Office of Tennessee
Nashville, Tennessee

Eugene Kiefer
Insurance Services Office of Oklahoma
Oklahoma City, Oklahoma

James Simms
Assistant Manager
Insurance Services Office of California
San Francisco, California

James Townley
Supervisor
Public Protection Department
Insurance Services Office of New Jersey
Parsippany, New Jersey

CONSULTANTS

Joseph Beckham
American Fire Equipment
Perry, Oklahoma

Samuel Cramer
Vice President
Aluminum Ladder Company
Florence, South Carolina

Jimmy Dickey
Chief, Plant Protection
International Harvester
East Moline, Illinois

Richard Engle
International Harvester
Rock Island, Illinois

Marie Haussermann
Burnham City Hospital
Champaign, Illinois

Frank A. McGurn, Jr.
Regional Executive Coordinator
National Automatic Sprinkler & Fire Control Association, Inc.
Crystal Lake, Illinois

C. H. McMillan
President
Fire Task Force Innovations, Inc.
Valparaiso, Indiana

William Matheny
Manager
Fleet Safety & Fire Protection
Gold Kist, Inc.
Atlanta, Georgia

Robert Noll
The General Electric Company
Lynn, Massachusetts

Robert Perkins
Safety Supervisor
National Cooperative Refinery Association
McPherson, Kansas

Charles Ray
Charles D. Ray & Associates
Albuquerque, New Mexico

Nick Renihan
Technical Promotions Manager
Akron Brass Company
Wooster, Ohio

Kenneth Stanton
Region Four Executive Coordinator
National Automatic Sprinkler & Fire Control Association, Inc.
Indianapolis, Indiana

W. S. Thompson
Chief Engineer
Elkhart Brass Manufacturing Company Inc.
Elkhart, Indiana

Steve Tinberg
Elkhart Brass Manufacturing Company, Inc.
Dallas, Texas

MANUAL PRICE LIST
PRICES AS OF JULY 1, 1979

101 FORCIBLE ENTRY, ROPE AND PORTABLE EXTINGUISHER PRACTICES
Types of forcible entry tools and general building construction; use of tools in opening doors, windows, roofs, floors, walls, partitions and ceilings; types, uses, and care of ropes, knots and portable fire extinguishers. 6th Edition - 1978 - $7.00.

102 FIRE SERVICE GROUND LADDER PRACTICES
Various terms applied to ladders; types, construction, maintenance, and testing of fire service ground ladders; detailed information on handling ground ladders and special tasks related to them. 7th Edition - 1973 - $6.00.

103 FIRE HOSE PRACTICES
Construction, care, and testing of hose and various fire hose accessories; preparation and manipulation of hose for rolls, folds, connections, carries, drags and special operations; loads and layouts for fire hose. 6th Edition - 1974 - $6.00.

104 SALVAGE AND OVERHAUL PRACTICES
Planning and preparing for salvage operations; care and preparation of equipment; methods of spreading and folding salvage covers; most effective way to handle water runoff; value of proper overhaul and equipment needed; determining cause of fire and recognizing and preserving arson evidence. 6th Edition - 1979 - $7.00.

105 FIRE STREAM PRACTICES
Characteristics, requirements and principles of fire streams; developing, computing, and applying various types streams to operational situations; formulas for application of hydraulics to the basic laws of science; actions and reactions created by applying streams under different circumstances. 5th Edition - 1972 - $6.00.

106 FIRE APPARATUS PRACTICES
Various types of fire apparatus classified by functions; driving and operating apparatus including pumpers, aerial ladders, and elevating platforms; maintenance and testing of apparatus. 5th Edition - 1970 - $6.00.

107 FIRE VENTILATION PRACTICES
Objectives and advantages of ventilation; requirements for burning, flammable liquid characteristics and products of combustion; phases of burning, backdrafts, and the transmission of heat; construction features to be considered; the ventilation process including evaluating and sizing up is discussed in length. 5th Edition - 1970 - $6.00.

108 FIRE SERVICE RESCUE AND PROTECTIVE BREATHING PRACTICES
Respiratory tract and respiratory hazards; protective breathing equipment and how to use, care, test, and inspect; other protection equipment including clothing, ambulances, and rescue equipment; rescue methods, techniques, procedures and situations. 4th Edition - 1969 - $3.50.

109 FIRE SERVICE FIRST AID PRACTICES
A completely revised edition of first aid; brief explanations of the nervous, skeletal, muscular, abdomen and digestive and genitourinary systems; injuries and treatment relating to each system; bleeding control and bandaging; artificial respiration, external cardiac compression and cardiopulmonary resuscitation; shock, poisoning, and emergencies due to heat and cold; fractures, sprains and dislocations; emergency childbirth; short distance transfer of patients and ambulances; conducting a primary and secondary survey; completely new narrative with numerous photographs to illustrate and complement narrative. 5th Edition - 1977 - $7.00.

110 FIRE PREVENTION AND INSPECTION PRACTICES
Fire prevention bureau and inspecting agencies; fire hazards and causes; prevention and inspection techniques; building construction, occupancy and fire load; special purpose inspections; inspection forms and check lists along with reference sources; maps and symbols; records and reports. 4th Edition - 1974 - $6.00.

200 ESSENTIALS OF FIRE FIGHTING
This new manual was prepared to meet the objectives set forth in levels I and II of NFPA, *Fire Fighter Professional Qualifications, 1974*. Included in the manual are the basics of: hose, ladders, fire streams, forcible entry, rescue, salvage and overhaul, ventilation, fire behavior science, ropes and knots, extinguishers, protective breathing, fire prevention, fire cause identification, ground cover, communications, water supplies and sprinkler systems. 1st Edition - 1977 - $15.00.

201 FIRE SERVICE PRACTICES FOR VOLUNTEER FIRE DEPARTMENTS
A general cursory introduction to material which is covered in detail in manuals 101 - 110 with the exception of First Aid. 5th Edition - 1971 - $6.00.

202 FIRE SERVICE ORIENTATION & INDOCTRINATION
History, traditions and organization of the fire service; operation of the fire department and responsibilities and duties of firefighters; fire department companies and their functions; a glossary of fire service terms and their definitions. 1st Edition - 1976 - $6.00.

203 FIRE SERVICE TRAINING PROGRAMS
Qualifications and duties of the training officer in planning, organizing, directing, and controlling the training program; examples of training centers, facilities, and programs; training curriculums — a master outline of hundreds of address sources. 2nd Edition - 1971 - $6.00.

204 PHOTOGRAPHY FOR THE FIRE SERVICE
Camera components and operations, films — their advantages and differences; fundamental principles in taking a good picture; processing the negative and print; controlling light for optimum results; accountability and storage of photographic materials; fire scene photography; using photographs as training aids; enhancing fire prevention and public relations through photography; how the investigator can utilize photographs; glossary of photographic terms. 1st Edition - 1976 - $6.00.

205 WATER SUPPLIES FOR FIRE PROTECTION
Importance, basic components, adequacy, reliability, and carrying capacity of water systems; specifications, installation, maintenance and distribution of fire hydrants; flow requirements, flow tests and control valves; sprinkler and standpipe systems. 3rd Edition - 1978 - $7.00.

206 AIRCRAFT FIRE PROTECTION AND RESCUE PROCEDURES
Aircraft types, engines, and systems; fire fighting apparatus (conventional and specialized), tools, clothing, extinguishing agents, dangerous materials, communications, pre-fire planning and airfield operations. 2nd Edition - 1978 - $7.00.

207 GROUND COVER FIRE FIGHTING PRACTICES
Ground cover fire apparatus, equipment, extinguishing agents, and fireground safety; organization and planning for ground cover fire; authority, jurisdiction, and mutual aid, techniques and procedures used for combating ground cover fire. 1st Edition - 1973 - $6.00.

208 RECORDS AND REPORTS FOR THE FIRE SERVICE
Procedures for compiling information for records; methods of writing and presenting reports; various sample forms for all activities of the fire service. 1st Edition - 1975 - $6.00.

209 FIREFIGHTER SAFETY
Basic concepts and philosophy of accident prevention; essentials of a safety program and training for safety; station house facility safety; hazards enroute and at the emergency scene; personal protective equipment; special hazards including chemicals, electricity, and radioactive materials; inspection safety; health considerations. 1st Edition - 1979 - $7.00.

210 PRIVATE FIRE PROTECTION & DETECTION
Automatic sprinkler systems, special extinguishing systems, standpipes, detection and alarm systems. Includes how to test sprinkler systems for the firefighter to meet NFPA 1001. 1st Edition - 1979 - $7.00.

301 THE FIRE DEPARTMENT OFFICER
The officers' functions analyzed; decisions, planning, activating, problem solving, delegating authority and supervision; fire fighting activities of fire officer. 2nd Edition - 1967 - $3.50.

302 FIRE DEPARTMENT — FACILITIES, PLANNING AND PROCEDURES
Fire protection facilities and water supplies; pre-fire planning and fire fighting procedures; dispatching companies, size-up, attack, confinement, extinguishment and overhaul. 2nd Edition - 1970 - $5.00.

303 FIRE SERVICE INSTRUCTOR TRAINING
Characteristics of good instructor; determining training requirements and what to teach; types, principles, and procedures of teaching and learning; training aids and devices; conference leadership. 3rd Edition - 1970 - $6.00.

304 FIRE PROBLEMS IN HIGH-RISE BUILDINGS
Locating, confining, and extinguishing fires; heat, smoke, fire gases, and life hazards; exposures, water supplies and communications; pre-fire planning, ventilation, salvage and overhaul; smokeproof stairways and problems of building design and maintenance; tactical check list. 1st Edition - 1976 - $6.00.

401 FUNDAMENTAL PRINCIPLES OF MATHEMATICS
Mathematics oriented to the fire service; fractions, decimals, measurements and weights, percentage and graphs, ratio, proportion, powers, and roots, algebraic and geometric fundamentals and problems. 3rd Edition - 1966 - $6.00.

402 FUNDAMENTAL PRINCIPLES OF SCIENCE
A basic investigation of matter, motion, force, machines, liquids, gases; principles of chemistry, combustion and heat, magnetism, electricity, atomic energy and radiation. 3rd Edition - 1967 - $6.00.

403 LEADERSHIP IN THE FIRE SERVICE
A series of lectures by Robert F. Hamm covering management, supervision, personnel administration, problem solving and analysis. 1st Edition - 1967 - $6.00.

500 FIREFIGHTER STUDY GUIDE
Correlates IFSTA publications with NFPA Standard 1001 to guide personnel through the requirements for national firefighter certification at levels I, II, III. Also provides a personal logbook for recording test dates, scores and instructor initials. 1st Edition - 1979 - $5.00.

502 FIRE SERVICE REFERENCE GUIDE
Each IFSTA manual, except 201 and 403, is individually subject indexed and cross-indexed by general subject; articles from *Fire Command, Fire Engineering, Fire Chief, Fire Journal, W.N.Y.F.* and *Fire News* indexed by subject 1970-76; index of FDIC Proceedings for six years. Separate 1978 *Addendum* indexes an additional year of reference material, plus subject index of IFSTA 101 and 206. 2nd Editions - 1977, 1978 - set for $6.00.

606 PUBLIC FIRE EDUCATION
A valuable contribution to your community's fire safety. Includes: public fire education planning, target audiences; seasonal fire problems; smoke detectors; working with the media; burn injuries; and resource exchange. 1st Edition - 1979 - $7.00.

INSTRUCTOR GUIDE SETS

Available for manuals 101 - 110, 206, and for "Fire Department Support of Automatic Sprinkler Systems." Basic plan for instructors to follow, tips to instructor; suggested supplemental publications and film references.

$2.00 per guide sheet — or $25.00 for set of twelve and free binder.

FIRE DEPARTMENT SUPPORT OF AUTOMATIC SPRINKLER SYSTEMS
The origin, development, and analysis of automatic sprinkler systems; water supply for; and fire department inspections and support of automatic sprinkler systems.

1st Edition - 1966 - $2.00 or free with a set of Sprinkler Slides.

SLIDES

Ladder	**$30.00**
***Salvage**	**75.00**
Sprinkler	**35.00**

2-inch by 2-inch slides that can be used in any 35 mm slide projector; supplements to respective manuals and sprinkler guide sets.

*The complete package consists of the slides, instructor's manual and instructor's guide sets.

Smoke Detectors Can Save Your Life	**50.00**
How To Inspect A Restaurant	**45.00**
Matches Aren't For Children	**20.00**
Public Relations for the Fire Service	**50.00**

Slide sets include narrative for complete presentation capabilities.

MANUAL BINDERS
Holds 3 - 10 manuals; polyflex, expandable pole binders. $6.00.

GUIDE SHEET BINDERS
Holds all 12 guide sets.
$5.00 or free with purchase of all 12 guide sets.

WATER FLOW TEST SUMMARY SHEETS
50 summary sheets and instructions on how to use; logarithmic scale to simplify the process of determining the available water in an area. $2.00.

PERSONNEL RECORD FOLDERS
Personnel record folders should be used by the training officer for each man on the department. Such data as IFSTA training, technical training (seminars), and college work can be recorded in this file, along with other valuable information. Letter size or legal size - $.40.

NFPA STANDARDS
1001 — "Fire Fighter Professional Qualifications, 1974." $4.00.

1002 — "Fire Apparatus Driver/Operator Professional Qualifications, 1976." $3.00.

1003 — "Airport Fire Fighter Professional Qualifications, 1978." $4.00.

1021 — "Fire Officer Professional Qualifications, 1976." $3.00.

1031 — "Fire Inspector, Fire Investigator and Fire Prevention Education Officer Professional Qualifications, 1977." $3.00.

1041 — "Fire Service Instructor Professional Qualifications, 1976." $3.00.

NFPA PUBLICATIONS

TXT-3	Management in the Fire Service, $16.50.
TXT-5	Introduction to Fire Protection Law, $14.95.
TXT-5A	Instructor's Manual (to above text), $3.50.
SPP-IC	Fire Protection Guide on Hazardous Materials, $12.50.
SPP-38	Arson: Some Problems and Solutions, $5.75.
FSP-7B	Fire Officer's Guide — Operating Aerial Ladders, $9.25.
FSP-38A	Fire Officer's Guide to Emergency Action - Revised Edition, $7.50.
FSP-36	Fire Officer's Guide to Dangerous Chemicals, $10.50.
FSP-43	Fire Officer's Guide to Company Leadership and Operations, $10.50.
FSP-27A	Fire Officer's Guide to Fire Apparatus Maintenance, $10.50.
FSME-1	Handling Personnel Problems at Company Level, $25.00.
FSP-34	How to Judge Your Fire Department, $4.95.
FSP-48	Fire Officer's Guide to Disaster Control, $16.50.
FPH-1476	Fire Protection Handbook, $43.50.
101	Life Safety Code, $5.00.

MANUALS

	IFSTA Number	Unit Price	No. of Each
Forcible Entry	101	7.00	
Ladder	102	6.00	
Hose	103	6.00	
Salvage & Overhaul	104	7.00	
Fire Streams	105	6.00	
Apparatus	106	6.00	
Ventilation	107	6.00	
Rescue & Protective Breathing	108	3.50	
First Aid	109	7.00	
Inspection	110	6.00	
Essentials	200	15.00	
Volunteer	201	6.00	
Indoctrination	202	6.00	
Training Programs	203	6.00	
Photography	204	6.00	
Water Supplies	205	7.00	
Aircraft	206	7.00	
Ground Cover	207	6.00	
Records & Reports	208	6.00	
Safety	209	7.00	
Private Fire Protection	210	7.00	
Fire Dept. Officer	301	3.50	
Facilities	302	5.00	
Instructor Training	303	6.00	
High-Rise	304	6.00	
Math	401	6.00	
Science	402	6.00	
Leadership	403	6.00	
Study Guide	500	5.00	
Reference Guide and Addendum	502	6.00	
Public Fire Education	606	7.00	

INSTRUCTOR'S GUIDE SETS

Forcible Entry	101	2.00	
Ladder	102	2.00	
Hose	103	2.00	
Salvage & Overhaul	104	2.00	
Fire Streams	105	2.00	
Apparatus	106	2.00	
Ventilation	107	2.00	
Rescue & Protective Breathing	108	2.00	
First Aid	109	2.00	
Inspection	110	2.00	
Aircraft	206	2.00	
Sprinkler		2.00	
Set with Binder		25.00	
IGS Binder Purchased Separately		5.00	

SLIDES

Ladders	30.00	
Salvage	75.00	
Sprinkler*	35.00	
Matches Aren't For Children	20.00	
How to Inspect a Restaurant	45.00	
Wake Up! Smoke Detectors Can Save Your Life	50.00	
Public Relations for the Fire Service	50.00	

*With purchase of Sprinkler Slides, a Guide Sheet is included free.

MISCELLANEOUS

Water Flow Test Sheets	2.00	
Personnel File Folders		
Letter	.40	
Legal	.40	
IFSTA Tie-Tac	2.50	
IFSTA Ladies' Pin	2.50	
IFSTA T-Shirt	5.00	
IFSTA Belt Buckle	5.00	
Manual Binders	**6.00**	

NFPA MATERIALS

(Selected list now available from IFSTA)

1001	(Firefighter)	4.00	
1002	(Driver)	3.00	
1003	(Airport)	4.00	
1021	(Officer)	3.00	
1031	(Inspector)	3.00	
1041	(Instructor)	3.00	

Management	16.50	
Fire Protection Law	14.95	
Law (Instructor's)	3.50	
Hazardous Materials	12.50	
Arson: Problems and Solutions	5.75	
Aerial Ladders	9.25	
Emergency Action	7.50	
Dangerous Chemicals	10.50	
Leadership and Operations	10.50	
Apparatus Maintenance	10.50	
Personnel Problems	25.00	
How to Judge Fire Department	4.95	
Disaster Control	16.50	
Fire Protection Handbook	43.50	
Life Safety Code	5.00	

ORDERING INFORMATION

- Send full payment with your order and IFSTA pays for postage and handling.
- If you choose to be billed for your order, IFSTA must add a shipping and handling charge based on the total dollar amount of your order as follows:

Order Size	Shipping Charge	Order Size	Shipping Charge
Under $20	add 10%	$100 to $200	add 5%
$20 to $99	add 7%	Over $200	add 3%

Payment Enclosed ☐

Bill Me Later, Plus Postage and Handling ☐

SUBTOTAL **$** ________

Discount, if applicable ________

Postage and Handling, if applicable ________

TOTAL **$** ________

Name ________

Phone # ________ **Dept.** ________

Address ________

City ________ **State** ________ **Zip** ________

No Postage
Necessary if
Mailed in the
United States

BUSINESS REPLY MAIL

FIRST CLASS PERMIT NO. 299 STILLWATER, OKLA.

POSTAGE WILL BE PAID BY ADDRESSEE

INTERNATIONAL FIRE SERVICE TRAINING ASSOCIATION

Fire Protection Publications
Oklahoma State University
Stillwater, Oklahoma 74074

Fold Here & Staple At Top